Introduction

When Sidney Camm's masterpiece, the Hawker Hurricane, entered RAF service in late 1937 it quickly became the most important aircraft in Britain's military arsenal and, arguably, remained so for the next three years. Bomber Command would have vehemently disagreed with this statement at that time, however, following the Munich Crisis, the Battle of France and the Battle of Britain and the struggle to introduce the Supermarine Spitfire in to service in quantity, it is, nonetheless, a statement that retrospectively, holds true.

A 1930s design, the Hawker Hurricane evolved through several versions and adaptations, and was one of the most successful aircraft types of the Second World War, one which served in every Theatre, from Norway and France, and the Battle of Britain – where its pilots scored more 'kills' than the rest of the British defences combined. Further afield, it fought in the defence of Malta, the campaigns in the Mediterranean, and North and East Africa; on the Russian Front and in the Far East where it fought against the Japanese over Singapore, the Dutch East Indies and Ceylon (modern day Sri Lanka) and saw widespread service throughout the Burma campaign until the end of hostilities.

Hurricanes were exported to Canada, South Africa, Belgium, Finland, Rumania, Portugal, Ireland, France, Netherlands East Indies, India, Turkey, Iran and Yugoslavia and under RAF command it also flew with Greek and Australian squadrons. Yet more were supplied to the Soviet Union under Lend-Lease agreements.

As well as a land-based interceptor fighter, fighter-bomber and ground support aircraft, the type was modified to enable it to operate from merchant ships, initially as catapult-launched convoy escorts aboard the famed CAM ships. They were followed in turn by fully navalised Sea Hurricanes that could operate off aircraft carriers, playing a major role in the vital Battle of the Atlantic and Arctic convoys, the invasions of Madagascar and French North Africa and experiencing its finest hour in protecting the crucial Operation *Pedestal* convoy to Malta during August 1942.

Acknowledgements

Of the many individuals who assisted in the compilation of this book we would like to take this opportunity to gratefully acknowledge the assistance of the following people: Mike Smith, Rosalyn Blackmore and all of the staff at the Newark Air Museum for their kind hospitality and access to the museum's archives; Phil Butler and Tony Buttler for the use of their photographs; Bill Newton; Steve Nichols for his superb colour illustrations, and, by no means least, to Tony O'Toole, not only for his excellent modelling skills but for his enthusiastic and unparalleled encyclopaedic knowledge of the Hurricane. Thank you all.

Martin Derry and Neil Robinson

An early production Hurricane IIa, Z2521, the first airframe from the fifth Hawker production run, comprising 1,000 Mk.IIs that were delivered from their Langley facility from 14th January 1941, photographed shortly after its completion. The aircraft is still in the Temperate Land Scheme and typical 1940 markings but without the Sky spinner and rear fuselage band introduced in the November, which were initially applied by the MUs before issue to an operational squadron. Z2521 served with Nos.249 and 242 Squadrons before joining No.247 Squadron circa June 1941 followed by No.235 Squadron prior to being sent to Russia in mid-January 1942. *T Buttler collection*

Origins
Design and Development

A line of No.111 Squadron Hurricane Is at Northolt in early 1938 finished in the recently introduced Temperate Land Scheme with yellow outlined roundels. Identifiable are L1550, L1560 and L1566. This was the first operational unit to be equipped with the new fighter and, as can be seen, unit markings have yet to be applied. The aircraft are fitted with Watts two-blade wooden propellers, original style 'kidney' exhaust manifolds, unarmoured windscreens, fuselage-mounted instrument venturis (instead of underwing pitots) and lack a rear fuselage ventral strake which was added to later production machines. To the rear of the image a line of Fairey Battles are in evidence, while towards the extreme left sits a Hawker Hart which was owned by the Hawker Company. *Via PH Butler*

At the time the Hurricane was being developed, RAF Fighter Command consisted of just thirteen frontline squadrons, equipped with either the Hawker Fury, Hawker Demon or the Bristol Bulldog – all biplanes with fixed-pitch wooden propellers and non-retractable undercarriages. Sydney Camm's design to meet F.7/30, the Hawker P.V.3, was essentially a scaled-up development of the Fury and was not amongst the proposals submitted to the Air Ministry as a government-sponsored prototype. After the rejection of the P.V.3 Camm started work on a cantilever monoplane with a fixed undercarriage armed with four machine guns and powered by a Rolls-Royce Goshawk engine. Detail drawings were finished in early 1934 but failed to impress the Air Ministry sufficiently for a prototype to be ordered. Camm's response was to further develop the design, introducing a retractable undercarriage and replacing the Goshawk with a new Rolls-Royce engine, the PV-12, later to become the Merlin. In August 1934, a 1/10th scale model was made and sent to the National Physical Laboratory at Teddington. A series of wind tunnel tests confirmed the aerodynamic qualities of the design, and in September Camm approached the Air Ministry again. This time the response was favourable, and a prototype of the 'Interceptor Monoplane' was ordered.

The prototype K5083

At around the same time, the Air Ministry issued Specification F.5/34 which called for a fighter aircraft to be armed with eight machine guns. However, by this time, work had progressed too far to immediately modify the planned four-gun installation and, by January 1935, a wooden mock-up had been finished and although a number of suggestions for detail changes were made, construction of the prototype was approved and a new specification, F.36/34, was written around the design. In July 1935, this specification was amended to include installation of eight machine guns. Work on the airframe was completed at the end of August 1935 and the aircraft components were taken to Brooklands, where Hawker had an assembly shed, and re-assembled. Ground testing and taxiing trials took place during late October and on 6 November 1935 the prototype took to the air for the first time at the hands of Hawker's chief test pilot, Flight Lieutenant (later Group Captain) P. W. S 'George' Bulman, who was assisted by two other pilots in subsequent flight testing, Philip Lucas, who flew some of the experimental test flights and John Hindmarsh who conducted the firm's production flight trials.

RAF trials of the aircraft at Martlesham Heath began in February 1936 and were favourable, the aircraft being easy to fly with good control responses and no apparent vices. Hawker's proposed type name 'Hurricane' was approved by the Air Ministry on 26 June 1936. Further testing however showed that the Hurricane had poor spin recovery characteristics, with all rudder authority being lost. The situation was resolved by the Royal Aircraft Establishment (RAE), who established that the problem was caused by a breakdown of airflow over the lower fuselage which could be cured by the addition of a small ventral strake and an extension to the bottom of the rudder. This improvement came too late to be incorporated in the first few production aircraft, but was introduced on the sixty-first airframe built and all subsequent aircraft.

Design

Although faster and more advanced than the RAF's current frontline biplane fighters, the Hurricane's design construction was already outdated when introduced. It used traditional Hawker construction techniques, with a Warren truss, box-girder, primary fuselage structure with high-tensile steel longerons and duralumin cross-bracing using mechanically fastened rather than welded joints, with wooden formers and stringers down the rear fuselage covered by doped linen. The cockpit was mounted high in the fuselage, creating a distinctive 'humpbacked' silhouette. Pilot access was aided by a retractable stirrup mounted below the trailing edge of the port wing. This was linked to a spring-loaded hinged flap which covered a handhold on the fuselage just behind the cockpit. When the flap was shut the footstep retracted into the fuselage.

Initially, the wing structure consisted of two steel spars with chordwise wooden formers, which like the tailplane, elevator, fin and rudder, was also fabric covered. However, in April 1939, a wing of all-metal construction with stressed duralumin skin was designed which permitted a greater diving speed some 80mph faster than the earlier fabric-covered wing. Introduced on the production line (in the second production batch around the N2328 serial number range), all subsequent Hurricanes were fitted with 'metal' wings. A few fabric-wing Hurricanes were still operational by the time of the Battle of Britain in the summer of 1940, but most had had their wings replaced with metal ones during servicing or after repair.

An advantage of the fabric covered steel-tube structure was that bullets and cannon shells could pass right through the wood and fabric without exploding. Even if one of the steel tubes was damaged, the repair work required was relatively simple, and could be done by squadron groundcrew. A damaged stressed skin structure, like that used by the Spitfire, required more specialised equipment to repair. The fabric covered steel-tube structure also permitted the assembly of Hurricanes relatively easily under 'field conditions', for example crated Hurricanes were assembled in West Africa then flown across the Sahara to the Middle East Theatre. Equally, to save space, some Royal Navy aircraft carriers carried their reserve Sea Hurricanes in a dismantled state which were slung up on the hangar bulkheads and deckhead for re-assembly when needed.

The prototype and early production Hurricanes were fitted with a Watts two-blade, fixed-pitch, wooden propeller. Since this was inefficient at low airspeeds, the aircraft required a long ground run to get airborne. Trials with a de Havilland two-pitch (coarse and fine) three-bladed metal Hamilton Standard propeller was found to reduce the take-off run from 1,230ft to 750ft, and Hurricanes began to receive the new propeller during April 1939. Later, at the beginning of 1940, Rotol developed a hydraulically operated, constant-speed propeller driving Jablo compressed-wood blades, which started to be fitted to Hurricanes around the time of the Battle of France.

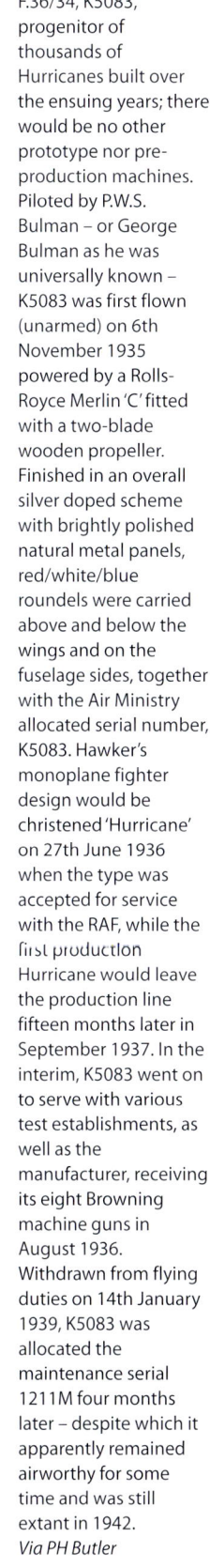

The solitary Hawker F.36/34, K5083, progenitor of thousands of Hurricanes built over the ensuing years; there would be no other prototype nor pre-production machines. Piloted by P.W.S. Bulman – or George Bulman as he was universally known – K5083 was first flown (unarmed) on 6th November 1935 powered by a Rolls-Royce Merlin 'C' fitted with a two-blade wooden propeller. Finished in an overall silver doped scheme with brightly polished natural metal panels, red/white/blue roundels were carried above and below the wings and on the fuselage sides, together with the Air Ministry allocated serial number, K5083. Hawker's monoplane fighter design would be christened 'Hurricane' on 27th June 1936 when the type was accepted for service with the RAF, while the first production Hurricane would leave the production line fifteen months later in September 1937. In the interim, K5083 went on to serve with various test establishments, as well as the manufacturer, receiving its eight Browning machine guns in August 1936. Withdrawn from flying duties on 14th January 1939, K5083 was allocated the maintenance serial 1211M four months later – despite which it apparently remained airworthy for some time and was still extant in 1942. Via PH Butler

With its ease of maintenance, widely-spaced main landing gear and benign flying characteristics, the Hurricane remained in use in operational theatres where reliability, easy handling and a stable gun platform were more important than outright performance, typically in roles like ground attack. One of the design requirements of the original specification was that the Hurricane was also to be used as a night fighter. Relatively simple to fly at night it shot down several enemy aircraft during the hours of darkness, while from early 1941, Hurricanes were used in the intruder role, patrolling German airfields in France, at night, in an attempt to catch night bombers when taking off or landing.

Production

The Hurricane was ordered into production in June 1936. War was looking increasingly likely, and time was of the essence in providing the RAF with an effective day fighter aircraft, it being unclear whether the admittedly more advanced Spitfire could enter production in time. As stated, the Hurricane's relatively simple construction using traditional manufacturing techniques was a boon, which was true for the operational squadrons as well, whose ground personnel were experienced in working on, and repairing, aircraft whose construction employed much the same principles as the Hurricane – the simplicity of its design enabling repairs to be made in squadron workshops. The Hurricane was also significantly quicker (and cheaper) than the Spitfire to build, requiring 10,300 man hours to produce rather than 15,200 for the Spitfire.

The first production order was for 600 Hurricanes with the serials L1547 to L2146, the first of which, L1547, powered by a 1,030hp Merlin II engine, first flew on 12 October 1937, while the first to enter RAF service joined No.111 Squadron at RAF Northolt in Middlesex two months later. By the outbreak of the Second World War nearly 500 Hurricanes had been produced equipping eighteen frontline squadrons.

Over 14,000 Hurricanes and Sea Hurricanes were eventually produced. Most (9,920) were built by the parent company, Hawker Aircraft Limited (HAL), initially at Brooklands in Surrey and then from 1941 at Kingston, Surrey, and Langley, then in Buckinghamshire, which produced them until 1944. Hawker's subsidiary, the Gloster Aircraft Company, purchased in 1934, manufactured 2,750 at Brockworth and Hucclecote in Gloucestershire. The Austin Aero Company at Cofton Hackett, a shadow factory of the Longbridge industrial complex at nearby Birmingham, built 300 and the Canadian Car and Foundry (CC&F) in Fort William, Ontario, Canada, produced over 1,400 Hurricanes.

During 1940, Lord Beaverbrook, the Minister of Aircraft Production, established an organisation in which a number of manufacturers were seconded to repair and overhaul battle-damaged Hurricanes. The Civilian Repair Organisation also over-

Despite the fact that this photograph was taken in poor lighting conditions it is of necessity included because this is L1547 – the very first production Hurricane from the very first production order which totalled 600 machines (serial range Ll547 to L2146). When photographed, this aircraft, was camouflaged in the Temperate Land Scheme with Dark Earth and Dark Green upper surfaces and Aluminium (silver painted) under surfaces with national markings in six positions. It was devoid of any other markings, not even a serial number at that moment in time. First flown on 12 October 1937, L1547 not unexpectedly spent a good deal of its life with a number of test establishments, although, by the time of its loss, on 10 October 1940, it was operational with No.312 Squadron when it was abandoned in flight over the River Mersey. *Newark Air Museum*

Head on view of L1547, showing the eight machine gun ports, more usually photographed with doped fabric patches over them. This image also illustrates to good effect both the thickness of the type's wing - which proved beneficial in allowing a close grouping of the two four-gun batteries to be obtained - and its original Watts two-blade, fixed-pitch propeller. The upper/under surface camouflage demarcation from the nose to the leading edges of the wing root is interesting and emphasises the curved nature of the fairing. *Via PH Butler*

hauled aircraft which were later sent to training units or to other air forces. One of the factories involved was the Austin Aero Company's Cofton Hackett plant and another was David Rosenfield Ltd, based at Barton aerodrome near Manchester.

Despite the RAF's urgent need for modern fighters, export orders were permitted and a number of Hurricanes were sold from the first production batches, with licences also being granted for the manufacture of the aircraft in Belgium and Yugoslavia – the details of these are included towards the end of this narrative.

Into RAF service

As mentioned, the first RAF unit to take delivery of the Hurricane was No.111 Squadron, at Northolt near London, which received its first four aircraft in late December 1937. Whilst training on their new mounts, the squadron also served as the RAF's 'service trials unit' thus easing the burden of the other fighter squadrons which would follow. Having previously flown fixed undercarriage, open-cockpit biplane fighters like the Bulldog, Fury, Gauntlet and Gladiator (albeit the latter featured an enclosed cockpit, flaps and four machine guns), the Hurricane presented a challenge to the pre-war pilots of Fighter Command, who nevertheless soon learnt to appreciate the new monoplane's power, speed, manoeuvrability and ruggedness.

Squadron Leader John Gillan, the CO of No.111 Squadron, established a record for their new mount on 10 February 1938 when he flew 327 miles from Edinburgh to London in 48 minutes at an average speed of 408.75mph. He was awarded the Air Force Cross for this feat as well as his leadership of the squadron during the introduction of the Hurricane into RAF service, but the record had in fact been aided by an 80mph tailwind, (which was officially downplayed at the time in order to impress the Germans) and thereafter Gillan became 'Downwind Gillan'.

Fifty Hurricanes had reached frontline operational squadrons by the middle of 1938. At that time, production was slightly greater than the RAF's capacity to introduce the new aircraft into service, hence the sale, with government approval, for the Hawker Company to sell the excess to nations likely to oppose German expansion.

Production was subsequently increased, with a plan to create a reserve of aircraft as well as re-equip existing and newly formed squadrons. Expansion Scheme E included a target of 500 fighters of all types by the start of 1938, yet by the time of the Munich Crisis, created by Germany's invasion of Czechoslovakia in September 1938, there were only two fully operational Hurricane squadrons of the planned twelve. A year later, when Germany invaded Poland, there were just eighteen operational Hurricane squadrons with three more in the process of converting.

Hurricane I, L1648, seen in 1938 whilst with No.85 Squadron, showing the disruptive camouflage scheme which was applied in two patterns, A Scheme and B Scheme, (B Scheme in this case). They were mirror images of each other and were applied on alternating airframes on the production line. Under surfaces were Aluminium (silver painted) at this time and the roundels on the wing upper surfaces and fuselage sides were outlined in yellow. The metal strap covering the metal wing root and fabric-covered outer wing panel join appears to still be in primer. This aircraft was damaged beyond repair at Debden on 6th October 1938. *T Buttler collection*

Hurricane Is, L1550 & L1559, No.111 Squadron, 1938, with minute unit badge's on their fins. This was the first squadron to receive the Hurricane, the first four of which arrived in December 1937 to reach their full establishment of sixteen in February 1938. This is a pre-Munich photograph, confirmed by the presence of L1550 which was written-off on 18th July 1938, while L1559 survived until January 1939. Later, this unit applied the numerals '111' to their fuselages for a period in 1938 which were presumably removed during the Munich crisis. These early Hurricanes featured Watts' two-bladed propellers, lacked a rear fuselage ventral strake and were finished in the Dark Green and Dark Earth upper surface camouflage scheme with overall Aluminium under surfaces. It is possible that the unit employed a 'training' Hurricane with yellow wing tips and a yellow band around the nose. *M Derry collection*

Above: Early Hurricane I, G-AFKX, formerly L1606 of Nos.151 and (probably) No.56 Squadrons which, having been despatched to the manufacturer for repair, was subsequently purchased by them and placed on the civil register. Unhindered by camouflage, military markings and radio mast, this familiar image captures the clean lines of a late-production Mark I, G-AFKX having been retrospectively modified by the Company in its new role as a trials and development aircraft. The obvious external alterations include the provision of the rear fuselage ventral strake, three-bladed Rotol propeller and metal wings – in fact this Hurricane was amongst the first to receive the latter and in all probability became the first to receive *production* metal wings as opposed to earlier, experimental, metal-clad wings. Much less obvious externally was the fact that this aircraft was also used to trial different variants of Merlin engine. G-AFKX's Form 113 shows that it was registered as such on 29 October 1938 and that the *registration* was withdrawn on 19 February 1946. However, this should not be taken as proof of G-AFKX's continuing existence in 1946, merely that that was when the record was updated for the last time! *Via PH Butler*

Below: Lord Nuffield greets test pilot R.C. Reynell in Hurricane I, L1791, seen in absolutely pristine condition with factory applied Night/White under surfaces albeit still with roundels and serial number. This aircraft is fitted with a de Havilland two-speed, three-blade metal propeller which replaced the original Watts two-bladed wooden propeller. L1791 was later allocated to No.46 Squadron which began receiving Hurricanes from early March 1939, thus allowing the replacement of their existing biplane Gauntlets IIs to commence. L1791 later served with No.7 OTU until it made a forced landing in July 1940 following engine failure. At the outbreak of war Australian-born Richard Reynell remained on secondment to Hawker but later joined No.43 Squadron on 27 August 1940. He was killed a few days later, on 7 September in Hurricane V7257, when his parachute failed to deploy after having been shot down by Bf 109s over London. *Newark Air Museum*

Into Battle

The Hurricane's baptism of fire came on 21 October 1939, when A Flight of No.46 Squadron took off from RAF Digby, Lincolnshire, and was directed to intercept a formation of nine Heinkel He 115D floatplanes from 1./KüFlGr 906, searching for ships to attack in the North Sea. The He 115s had already been attacked and damaged by two No.72 Squadron Spitfires when the six No.46 Squadron Hurricanes intercepted the Heinkels which were flying at sea level in an attempt to avoid further attacks. Nevertheless the Hurricanes shot down three of them in rapid succession and damaged another (although No.46 claimed five and No.72 claimed two!)

By late 1939/1940, many of the early delivery machines were in the process of being updated with 'metal' wings, 1,030hp Merlin III engines, ejector exhaust manifolds, de Havilland and Rotol variable speed three-blade propellers, reflector gunsights instead of the original ring and bead type, internal and external armoured windscreens and armour-plated rear cockpit bulkheads – none of which could be achieved overnight of course – resulting in a range of modifications, for a while, that numbered an estimated twenty-seven different standards.

The Phoney War

In response to a request from the French government for ten fighter squadrons to provide air support, in addition to ten squadrons of Fairey Battles that were flown to bases in metropolitan France in late August/early September 1939, Air Chief Marshal Sir Hugh Dowding, Commander-in-Chief of RAF Fighter Command, argued that this number of fighters would severely deplete Fighter Command's British defences, and so initially only a token force of four Hurricane squadrons, Nos.1, 73, 85 and 87, were sent to France in early September 1939, (all Spitfires being retained for Home defence). The RAF supplied two air contingents initially – the Advanced Air Striking Force (AASF) and the Air Compo-

Hurricane I, N2358 'Z' No.67 Wing (which formed on 6 November 1939), AASF, France, in early 1940. Operated by No.73 Squadron when photographed, N2358 is seen minus the squadron code 'TP'. In fact both AASF Hurricane squadrons (the other was No.1 Squadron) deleted their unit codes, or at least were meant too, to better match *Armée de l'Air* practice with whom they were closely operating, they even introduced French-style rudder stripes, albeit with the red stripe leading, for easier recognition by French forces. Fitted with a de Havilland two-speed metal propeller, this Hurricane wears Night/White under surfaces with underwing roundels, another recognition expedient applied to all RAF aircraft operating from or over the French mainland. N2358 arrived with the squadron in December 1939 having previously been allocated to No.43 Squadron, however, by 2 February it was with Glosters prior to be being despatched to Finland. *Crown via PH Butler*

A familiar image of two Hurricane Is, including L2001 'JU-B', of No.111 Squadron refuelling at Wick in early 1940. Carrying red/white/blue fuselage roundels, unusually L2001 is devoid of a roundel on its starboard wing upper surfaces, which may indicate it had recently been fitted with a replacement metal wing. Both aircraft are fitted with de Havilland two-speed propellers and pole-style radio aerials and feature Night/White under surfaces (without underwing roundels) but with nose, rear fuselage and tailplane under surfaces in painted Aluminium, a frequent variation of the Night/White under surface scheme. Having previously served with No.56 Squadron, L2001 served with No.111 Squadron until it was destroyed on 19 June 1940 following engine failure whilst taking off. *Crown via PH Butler*

nent of the British Expeditionary Force (BEF). The four Hurricane squadrons initially formed No.60 Wing within the Air Component of the BEF, but by the middle of September further RAF squadrons comprising Blenheim IV bombers and Lysander tactical reconnaissance and army co-operation aircraft started arriving. Over the following autumn and winter, the squadrons were rotated around various bases while Nos.1 and 73 Squadrons were detached from the BEF's Air Component control during the winter to form No.67 Fighter Wing attached directly to the AASF.

On 30 October, Hurricane pilots experienced their first action over France. Pilot Officer P. W. O 'Boy' Mould of No.1 Squadron, flying L1842, shot down a Dornier Do 17P from 2.(F)/123, sent to photograph allied airfields close to the border, about 10 miles west of Toul, becoming the first RAF pilot to down an enemy aircraft on the continent in the Second World War. Flying Officer E. J. 'Cobber' Kain, a New Zealander, was responsible for No.73 Squadron's first victory, on 8 November 1939, whilst stationed at Rouvres. He went on to become one of the RAF's first 'aces' of the war, being credited with sixteen 'kills' before his death in a flying accident on 6 June 1940.

Hurricanes were also involved in the German invasion of Norway. On 9 April 1940, under codename Operation *Weserübung* the *Wehrmacht* invaded Denmark, which capitulated after a day, but Norway continued to resist. On 14 April Allied ground troops were landed in Norway, but by the end of the month, the southern parts of the country were in German hands. On 14 May 1940, No.46 Squadron embarked on HMS *Glorious* and sailed for an airfield near Harstad, Norway, to augment the Gladiators of No.263 Squadron operating from improvised airfields and the frozen lake at Lesjaskog, but they had to return with the carrier to Scapa Flow when the landing ground was found to be unusable.

On 26 May, ten of the squadron's Hurricanes were flown off to Skaanland, but due to the soft surface two crashed on landing so the remainder were diverted to Bardufoss, sixty miles further north. After providing fighter cover for the Narvik area for two weeks the order to evacuate all Allied forces from Norway was received and, on 7 June, despite the lack of arrester hooks and no deck landing training, the squadron flew its surviving Hurricanes back on to *Glorious'* deck – all landing safely. The squadron's ground crews embarked in other ships and re-assembled at Digby, though tragically, HMS *Glorious* and her destroyer escort were intercepted by the German battleships *Scharnhorst* and *Gneisenau* during their return home, and sunk. Only two RAF officers survived the sinking, one being No.46's CO, Squadron Leader K. B. B. (later Air Chief Marshal Sir Kenneth) Cross. Despite this disaster the squadron was operational again by the end of June, at Digby.

Battle of France

By the spring of 1940, it became rapidly apparent that the handful of Hurricane squadrons based in France would be woefully inadequate to offset an impending *Luftwaffe* avalanche. In May, three more Hurricane squadrons, Nos.3, 79 and 504, were sent to reinforce the earlier units as Germany's *Blitzkrieg* gathered momentum. On 10 May, the first day of the Battle of France, Hurricane squadrons claimed forty-two *Luftwaffe* aircraft shot down for the loss of seven Hurricanes with none of the pilots killed. Hurricane units also escorted bombers, including those involved with the raids against the Vroenhoven and Veldwezelt bridges on the Meuse, at Maastricht by No.12 Squadron's Fairey Battles on 12 May. The escort consisted of eight Hurricanes from No.1 Squadron, but when the formation approached Maastricht, it was bounced by Bf 109Es from 2./JG 27. Two Battles and

two Hurricanes were shot down, two more Battles were brought down by flak and the fifth was forced to crash land.

On 13 May 1940, more Hurricanes arrived, bringing the total of Hurricane squadrons operating from French soil to ten – Nos.1, 3, 73, 79, 85, 87, 242, 501, 504 and 615 Squadrons (No.615 having exchanged its Gladiators for Hurricanes in the preceding weeks) – but heavy losses continued and by the end of the first week of fighting only three of the squadrons remained near operational strength. With ferocious air combat continuing from dawn to dusk, throughout May, the order was finally received on the afternoon of 20 May 1940 for all Hurricane units based in northern France to abandon their bases and return to the UK. During eleven days of fighting in France, between 10 to 21 May, Hurricane units claimed 499 'kills' and 123 probables, although contemporary German records examined post-war, attribute 299 *Luftwaffe* aircraft destroyed and sixty-five seriously damaged by RAF fighters. Number 1 Squadron was the most successful of the campaign, claiming sixty-three victories for the loss of five pilots. On the evening of 21 May, the only Hurricanes still operating in France were those of the AASF that had been moved to bases around Troyes and when the last Hurricanes left France, of the 452 Hurricanes sent only sixty-six returned to bases in the UK with over 170 having to be abandoned at their airfields.

During Operation *Dynamo* – the evacuation of British, French and Belgian troops cut off by the German army surrounding Dunkirk – Hurricanes continued to operate from British bases and it was over Dunkirk that the *Luftwaffe* suffered its first serious rebuff of the war. Although operating from captured bases in France, the Bf 109 was at the outer limits of its range and possessed less flying time over the area than the defending Hurricanes (and Spitfires) operating from airfields in southern England. *Luftwaffe* bombers, many still based in western Germany with farther to fly, found that British fighter attacks often prevented them from performing to their customary, often uninterrupted, degree of effectiveness and both sides suffered heavy losses, which for the *Luftwaffe*, came as a bit of a shock. For instance, *Fliegerkorps* II reported in its War Diary that it lost more aircraft on 27 May attacking the evacuation area than it had lost in the previous ten days of the campaign.

Initial engagements with the *Luftwaffe* had showed the Hurricane to be a tight-turning and steady platform but the Watts two-bladed propeller was clearly unsuitable and its replacement with de Havilland and Rotol units was a priority. The Merlin III engine was designed to run on standard 87 octane aviation fuel, but from early 1940, increasing quantities of 100 octane fuel became available, which together with modifications to allow an additional 6psi of supercharger boost for five minutes, increased engine output by nearly 250hp and gave the Hurricane an approximate increase in speed of 25 to 35mph below 15,000ft, which greatly increased the aircraft's climb rate. This form of emergency power was an important modification that allowed the Hurricane to be more competitive against the Bf 109E and to increase its margin of superiority over the Bf 110C, especially at lower altitudes.

Deliveries of new Hurricanes fitted with Rotol constant-speed propeller units (CSU) commenced in April/May 1940 and Hurricanes already in France were being retrofitted with Rotol CSUs by parties of the manufacturer's engineers flying out from England to do the work. The Rotol CSU transformed the Hurricane's performance and prompted de Havilland to undertake a modification programme of upgrading its older two-pitch propeller into a similar CSU, so that by the late spring/early summer of 1940, most frontline operational Hurricanes were fitted with either Rotol or de Havilland constant-speed propeller units.

The Battle of Britain

By the end of June 1940, following the fall and surrender of France on the 22nd, almost half of the RAF's Fighter Command squadrons were equipped with Hurricanes. A short lull ensued whilst the *Luftwaffe* replaced its losses from the French Campaign and established itself on the airfields in the countries they had captured. In Britain this time was spent in putting as many new fighters and trained pilots into service as possible to prepare against the attack everyone knew was coming. The future of Britain was about to be decided in the skies above southeast England, and, as the country's new Prime Minister, Winston Churchill, who took over the premiership on 10 May, put it, 'What General Weygand called the Battle of France is over, the Battle of Britain is about to begin'

The Battle of Britain officially lasted from 10 July until 31 October 1940, with the heaviest fighting taking place between the beginning of August and mid-September. On 16 July, Hitler ordered the preparation of a plan to invade Britain, under 'Directive No 16: The Preparation of a Landing Operation against England' better known today as Operation *Sealion*. All preparations were to be made by mid-August and it was scheduled to take place in mid-September 1940. *Sealion* called for landings on the south coast of England, backed by an airborne assault. Neither Hitler nor the *Oberkommando der Wehrmacht* (OKW, Supreme Command of the Armed Forces), believed it would be possible to carry out a successful amphibious assault on Britain until the RAF had been neutralised. It was believed that air superiority might make a successful landing possible although it would still be a very risky operation requiring absolute mastery over the Channel by the *Luftwaffe*.

Flight Sergeant Geoffrey 'Sammy' Allard getting out of his No.85 Squadron Hurricane I, probably during the Battle of Britain. Allard had previously seen action in the Battle of France, claiming eight 'kills' before the unit returned to the UK where he claimed another eight between 17 August and 1 September 1940. By the summer of 1940, most of the unit's Hurricanes had been fitted with constant-speed Rotol propeller units, which improved the aircraft's take-off run and climb-to-height rate. They had also acquired 'light coloured' under surfaces, applied at unit level, and displaying somewhat non-standard demarcations, in lieu of the official shade of Sky that was promulgated, but initially in very short supply, to replace the Night/White under surface scheme. In October the squadron was withdrawn to become a night-fighter unit, although even prior to October some aircraft had been fitted with anti-glare exhaust shields forward of the cockpit. Allard had been promoted to Flight Lieutenant, and awarded the DFC, DFM & Bar with a score of nineteen victories, when he was tragically killed in a flying accident on 13 March 1941.
Newark Air Museum

The Battle went through a series of phases:

Phase 1: From 10 July to 11 August 1940, which saw a series of running fights over convoys in the English Channel and occasional attacks on coastal shipping, convoys and harbours, such as Portsmouth, by Junkers Ju 87 *Stuka* dive bombers.

Phase 2: From 12 to 23 August 1940 when the *Luftwaffe* started to shift its attacks over to RAF airfields, the ground infrastructure and aircraft factories

Phase 3: Which saw intensified *Luftwaffe* attacks on RAF airfields from 24 August to 6 September 1940 – and came very close to destroying Fighter Command and its bases.

Phase 4: From 7 September to 31 October 1940, when the *Luftwaffe* changed its tactics and resorted to attacking areas of political significance such as London in daylight, using area bombing tactics.

Phase 5: From late September 1940 through to the spring of 1941 when the *Luftwaffe* turned more and more to a night bombing campaign against London and the UK's major cities – known as 'The Blitz'.

As may be imagined, with Hurricanes making up half of Fighter Command's frontline force, the type was heavily committed to the Battle and Hurricane squadrons were involved in all the Phases, including some of the first nocturnal interceptions when the *Luftwaffe* started night bombing raids from late September. Despite the undoubted abilities of the Spitfire, it was the Hurricane that scored the higher number of victories during this period, accounting for almost 60 per cent of the recorded 2,739 German losses. Although the Hurricane was slower than both the Spitfire and the Messerschmitt Bf 109E, with its thick wings which affected rapid acceleration, it could out-turn both of them. The Hurricane was a steady gun platform, and in spite of its performance differences compared to the Bf 109, the Hurricane was still a capable fighter, especially at lower altitudes. One tactic of the Bf 109 was to attempt to 'bounce' RAF fighters in a dive. If spotted in time, Hurricanes were able to evade such tactics by turning into the attack or going into a 'corkscrew dive', which the '109s, with their lower rate of roll, found hard to counter. If a Bf 109 was engaged in a 'straight dogfight', the Hurricane was just as capable of out-turning it as the Spitfire, although in a stern chase, the Bf 109 could easily outpace and evade the Hurricane.

In the summer of 1940, Hurricane Is, (and Spitfire Is) were powered by Merlin III engines, fitted with a float chamber SU carburettor. When a Hurricane (or Spitfire) performed a negative-G manoeuvre (i.e. pitching the nose hard down), fuel was forced up to the top of the carburettor's float chamber rather than into the engine, leading to loss of power. If the negative-G continued, then enough fuel would collect in the top of the float chamber to force the float to the floor of the chamber. This would in turn open a needle valve to maximum, flooding the carburettor and drowning the supercharger with an over-rich mixture which would lead to a cut-out, thus shutting down the engine completely – a serious drawback in combat!

Bf 109s and Bf 110s used Daimler-Benz DB 601 inverted V12 engines fitted with fuel injection pumps, not carburettors, which kept their fuel at a constant pressure whatever manoeuvres they performed and did not suffer from this problem. They could exploit the difference by pitching steeply forward whilst pushing the throttle wide open; pursuing British fighters were left 'flat footed' as trying to emulate the manoeuvre would result in loss of power, or fuel flooding and engine shutdown. The only British countermeasure available was to half-roll, so the aircraft would only be subjected to positive-G as they followed a German aircraft into a dive, which invariably took just enough time to let the enemy escape.

Complaints from the pilots led Beatrice 'Tilly' Shilling, a young engineer working at the Royal Aircraft Establishment (RAE) at Farnborough, to devise a disarmingly simple solution – a flow restrictor which was a small metal disc, much like a metal washer. The restrictor orifice was made to accommodate just the fuel needed for maximum engine power, the power setting usually used during dogfights. Whilst not completely solving the problem, the restrictor, along with modifications to the needle valve, permitted Hurricane and Spitfire pilots to perform quick negative-G manoeuvres without loss of engine power. In early 1941, Miss Shilling and a small team from the RAE travelled around Fighter Command's airfields fitting these restrictors, giving priority to front-line units and by March 1941 the device had been installed throughout RAF Fighter Command. Officially named the 'RAE restrictor', the device was immensely popular with pilots, who affectionately named it 'Miss Shilling's orifice' or simply the 'Tilly orifice'. This simple solution was only ever a stopgap as it did not allow inverted flight for any length of time, however, the problems were ultimately overcome by the introduction of Bendix and later Rolls-Royce pressure-carburettors in 1943.

Whilst the Hurricane was rugged, sturdy and stable, another problem was the gravity fuel tank, situated in the forward fuselage, right in front of the instrument panel, without any form of protection for the pilot. Many Hurricane pilots were seriously burned as a consequence of flames burning through the instrument panel. This became of such concern that Hawker retrofitted the fuselage tanks with a self-expanding rubber coating called Linatex. If the tank happened to be punctured by a bullet, the Linatex coating would expand when soaked with petrol and seal it. As if to illustrate this point, mention can be made here of the only Victoria Cross awarded to a member of Fighter Command

Another No.85 Squadron Hurricane I, P2923 'VY-R', photographed during the Battle of Britain, and regularly flown by another high scoring squadron pilot, (then) Pilot Officer Albert Gerald 'Zulu' Lewis, a South African, who had previously flown with Nos.616 and 504 Squadrons. He joined No.85 in France in April 1940 and on 19 May shot down five enemy aircraft over Lille before he was himself shot down. He was awarded the DFC on 18 August, the same day that P2923 went missing over the North Sea. Finished in typical Battle of Britain scheme and markings, it also featured non-standard 'light coloured' under surfaces applied higher up the nose than usual, with two yellow (?) rings around its Night-painted Rotol propeller spinner. Albert Lewis finished the war as a squadron leader adding a bar to his DFC, with a final tally of eighteen victories. *Newark Air Museum*

An undated image of unidentified No.85 Squadron Hurricane I, coded 'VY-Q' (possibly P3854) almost certainly taken during the Battle of Britain period. The aircraft is fitted with the later style radio aerial mast and a rear-view mirror and has correctly dimensioned fuselage roundels. *Newark Air Museum*

during the war, which occurred during the Battle of Britain. On 16 August 1940, Flight Lieutenant James Brindley Nicolson was leading a section of three No.249 Squadron Hurricanes over Southampton when they were 'bounced' by Messerschmitt Bf 110s. All three Hurricanes were hit and Nicolson was wounded in one eye and a foot. His Hurricane's gravity tank was set alight engulfing the cockpit in flames. About to abandon his aircraft, Nicolson noticed that one of the Bf 110s had overshot his aircraft, so he remained in the cockpit, by now an inferno, fired at the enemy and then bailed out. Despite serious burns to his hands, face and legs, he managed to open his parachute in time, only to be fired on by a member of the Home Guard on his descent!

Even before the Battle of Britain officially commenced, Hurricanes were sent to fight elsewhere. On 10 June 1940, Italy, rushing to the aid of the German victor, declared war on France and Britain. Mussolini immediately took advantage of Britain's preoccupation with its forthcoming battle to survive and called for an offensive throughout the Mediterranean. Within hours, the first bombs were dropped on Malta, a strategically significant British Crown Colony located between Gibraltar and Alexandria, a fortress island vital to Britain's lifeline to Egypt (which Italy attacked in September), Suez, India and the Far East.

At the start of the *Regia Aeronautica*'s aerial bombardment of Malta, the only fighter defence comprised a small number of ex-Fleet Air Arm Gloster Sea Gladiators which together formed the Hal Far Fighter Flight.

However, the Flight was soon augmented by a few Hurricanes that had been flown across France prior to the latter's surrender. All of the Hurricanes sent via this route were intended for the defence of Egypt, using Malta as a staging post, but, by dint of appeal, a total of five had been retained in Malta by 24 June (with six others continuing their journey to Egypt) to augment the Sea Gladiators. Of course there were no spare parts for Malta's Hurricanes and by the end of June they were all grounded. Consequently, a dozen Hurricane Is were collected in the UK as No.418 Flight RAF which embarked aboard the small carrier HMS *Argus* in late July 1940 for urgent delivery to beleaguered Malta under Operation *Hurry*. By 2 August, *Argus* was southwest of Sardinia and all twelve Hurricanes were successfully launched. All reached Malta where they ultimately formed part of a new No.261 Squadron. This was the first in a long series of similar operations to ferry fighter aircraft to Malta using aircraft carriers, although the next attempt, Operation *White*, in November 1940, ended in disaster – just four of the twelve Hurricanes despatched arrived at Malta.

On 11 November, *Argus* again sailed from Liverpool with a deck load of another dozen Hurricanes for delivery to Malta (Operation *White*). However, due to a perceived threat of interception by the Italian Fleet it was decided to launch the Hurricanes when still 350 nautical miles west of Malta, the British admiral having been advised that the Hurricanes could safely cover 400 miles – within the ferrying range of both the Hurricanes and their escorting Skuas. Therefore, before dawn on 17 November, the first flight of six Hurricanes took off from *Argus* at 06:15hrs.

Given the correct speed and the best cruise range, the Hurricanes would have been left with just 45 minutes of fuel after reaching Malta, but they used up a lot of fuel whilst forming up, then the fighters flew at 150mph at 2,000ft, far from the ideal height and speed intended for their maximal range. The second wave was launched an hour later, as the convoy turned back at full speed. Unfortunately the wind veered from southwest to southeast, hampering the eastward path of the aircraft. A Sunderland flying boat met the first wave to lead them to Malta, even so two Hurricanes were lost after running out of fuel (one of the pilots was rescued by the Sunderland), while the four

Another unidentified Hurricane I belonging to No.85 Squadron; probably taken during the Battle of Britain period. Although this image is believed to have been taken for publicity purposes, to judge by the armourer's labours it doesn't appear to be so obviously 'staged' as so many other wartime photographs were. The object painted below the cockpit coaming is not the unit's hexagon motif, it is in fact a partially obscured Native American Indian war bonnet. Their hexagon motif had been applied to the unit's Hurricanes' fins whilst in France, discarded by the time the Battle of Britain commenced, only to be reintroduced, probably at the instigation of their CO, Sqn Ldr Peter Townsend and reapplied below the port-side cockpit coaming.
Newark Air Museum

remaining Hurricanes and their accompanying Skua landed at Luqa at 09:20hrs.

Unfortunately the second flight were denied their Sunderland escort because it failed to take off from Gibraltar. They also missed a bomber sent from Malta to replace the Sunderland. All of the Hurricanes ran out of fuel and crashed into the sea, with the loss of all the pilots, only the accompanying Skua survived to crash-land in Sicily just before it too ran out of fuel.

Following the fall of France, and given the Royal Navy's shortage of aircraft carriers, an alternate method of supply was pioneered in order to provide Egypt with more aircraft. The method chosen was the Takoradi Route. Situated on the Gold Coast (a British colony on the Gulf of Guinea in West Africa which became Ghana in 1957), crated aircraft, including Hurricanes, were delivered by sea to Takoradi, assembled and test flown there, then ferried across Africa to Khartoum – a route first pioneered by Air Vice-Marshal Arthur Coningham in 1925. The Takoradi Route to Egypt was first used by RAF aircraft

Hurricane I, P2798, 'LK-A', No.87 Squadron, Church Fenton, Yorkshire, August/September 1940 with Flt Lt Ian 'Widge' Gleed, DFC, in the cockpit. Finished in the standard Temperate Land Scheme of Dark Earth and Dark Green upper surfaces, to the B Scheme pattern with 'Sky' under surfaces, P2798 had a red spinner when flown by Gleed who was the Squadron's A Flight Commander. National markings appear to have been applied using the pre-war 'bright shades' of red and blue on the fuselage roundels, whilst the yellow outer ring was truncated along the lower demarcation line. 'Figaro' the cat, a cartoon character from Walt Disney's 'Pinnochio' film, and Gleed's personal marking, was applied to the emergency break-out panel door on the starboard side of the aircraft. P2798 formed part of the first Gloster-built production order for 500 Hurricanes, deliveries of which commenced in November 1939 and were completed five months later. Fitted with a Rotol constant speed propeller, P2798, albeit with serial number overpainted, served only with No.87 Squadron until it was abandoned in flight following engine failure in October 1941. *Tony O'Toole collection*

Hurricane I, V7434 'DZ-R', No.151 Squadron photographed towards the end of the Battle of Britain period and featuring standard colour scheme and markings with factory-applied fuselage roundels and fin flashes of the correct diameters and dimensions and almost certainly finished in the correct shade of Sky on the under surfaces. The marking on the fuselage spine just above the fuselage roundel is thought to be a Maori god or good luck symbol, possibly indicating the pilot was a New Zealander. V7434 exhibits all the features of an airframe that is fully up to late 1940 standards, including a constant-speed Rotol propeller, armoured windscreen and VHF T/R Type 1133 radio aerial mast. This machine only ever served with No.151 Squadron, surviving until October 1940 when it crashed at Digby. *Newark Air Museum*

Opposite page:
Epitomising the Hurricane in its new offensive role from the middle of 1941 in the European Theatre of Operations, 'taking the fight back to the enemy', Hurricane IIb, BE485 'AE-W' of No.402 Squadron (previously No.2 Squadron RCAF). Originally equipped with the Hurricane I until replaced by the Mk.IIa from May 1941, the latter Mark was replaced by the Mk.IIb two to three months later. Number 402 Squadron helped to pioneer operations with bomb-carrying Hurricanes and in November 1941 eight of the unit's 'Hurribombers' - each carrying two 250lb bombs - flew their first bombing mission when they attacked an airfield in France. BE485, seen here with two such bombs, subsequently served with No.175 Squadron until forced to ditch in the Channel in April 1942 when returning from an operational sortie.
T Buttler collection

in September 1940, when one Blenheim guided six Hurricanes to their destination.

By the end of November 1940, the RAF in Egypt had been bolstered by sufficient Hurricanes to equip Nos.73 and 274 Squadrons, ready for the beginning of Operation *Compass*, the first major British military operation of the Western Desert Campaign in the Second World War, when, from December 1940 to February 1941, British and Commonwealth forces attacked the Italians in western Egypt and eastern Libya. The operation was a complete success with the Hurricanes, Blenheims and Lysanders, finding it hard to keep pace with the ground forces, often landing after a sortie at a more advanced strip than the one from which they had set out.

Hurricane II

In the autumn of 1940, the Hurricane IIa Series I started entering service, albeit in small numbers at first. Powered by the 1,480hp Merlin XX, which utilised a two-speed supercharger resulting in increased power at higher altitudes, it was designed from the outset to run on 100 octane fuel which allowed higher manifold pressures – achieved by increasing the boost from the centrifugal supercharger – giving the Mk.II a maximum speed of 340mph. Another improvement, was the use of a 70/30 per cent water/glycol coolant mix rather than the 100 per cent glycol of earlier versions. This substantially improved engine life and reliability, removed the fire hazard of the inflammable ethylene glycol, and reduced the oil leaks that had been a problem with the early Merlin I, II and III series of engines.

The new engine was longer than earlier Merlins and so the Hurricane gained a 4½ inch 'plug' in front of the cockpit firewall, which had the added benefit of making the aircraft slightly more stable due to the slight forward shift in its centre of gravity. The Rotol CSU unit was now housed within a specially designed, more aerodynamically pointed, propeller spinner, the original 'blunt' spinner having originally been designed for the Spitfire which possessed a slightly greater diameter nose contour than the Hurricane. Additionally, Dowty levered-suspension units with a 'knuckled' torque-link tailwheel leg bagan to replace the straight-legged internally sprung Hawker/Lockheed design introduced on the Mk.I.

Well before the outbreak of war, it was realised that the increasing speeds and robustness of aircraft would call for something more powerful than the .303inch rifle-calibre machine gun, eight of which were fitted to the Hurricane and Spitfire at the time. The rapid introduction of armour protection during 1940 greatly reduced the effectiveness of rifle-calibre guns, especially against bombers which, whilst being easier to hit, were more difficult to destroy. Many German bombers returned to base despite being riddled by .303 bullets, protected as they often were by crew seat armour and self-sealing fuel tanks. Increasing the number of .303 machine guns to twelve in the Hurricane IIb was one attempt to achieve a greater weight of fire, but a bigger gun was needed.

Extensive tests by Browning and Vickers of .50 calibre heavy machine guns and 20mm cannon in the 1920s had provided important information, but the .50s were rejected: not only were they much heavier and slower firing than the .303s, they did not have the benefit of an effective high-explosive shell. Hence the RAF decided to look for a 20mm cannon and, in 1935, the Air Staff ordered a French Dewoitine 510 Fighter equipped with a single engine-mounted Hispano Suiza 20mm *Moteur Cannon*. The D.510 and its gun, which outclassed any similar weapon on the market at this time, commenced flight trials at Martlesham Heath in 1937.

Thereafter a significant effort was made to get the Hispano 20mm cannon into service, but despite its high priority, the problems associated with acquiring manufacturing rights, redesigning the gun's feed mechanism and setting up a suitable manufacturing process delayed matters. Further, problems would be experienced in adapting the gun to fit within the wings of both the Hurricane and Spitfire, it often being overlooked that the *Moteur Cannon* was designed to sit within the rigid, non-flexing, environment of an engine cylinder block. (This fact also accounted for the gun's length as the barrel had to clear the propeller boss. Much later, when the Hispano V was being developed to replace the Mk.II it was discovered that 12 inches could be lopped off of the barrel without materially affecting its ballistic properties.)

Number 19 Squadron, was equipped with modified Spitfire Is that had been fitted with re-built wings, each armed with a 20mm Hispano, and tested in June/July 1940. The results were disappointing; the gun did not respond well to being mounted on its side (in order to enclose the drum magazine within the Spitfire's thin wing), nor to being installed in a wing that flexed in flight. Reliability was so poor and stoppages so frequent, that the squadron asked for its .303-armed Spitfires back.

Hawker had also experimented with improving the armament of its fighter by fitting cannon. The first trials used the 20mm Hispano I – one fitted beneath each wing in a streamlined fairing – on Hurricane I, L1750, which was subsequently allocated to No.151 Squadron at North Weald during the Battle of Britain. However, the Hurricane's thicker wing section allowed for a more reliable and less drag-inducing installation to be made and together with small changes to the feed mechanism and cannon mountings, four 20mm Hispano II cannon, two per wing, were successfully fitted, although the additional weight did reduce performance. Small clearance blisters were also needed on the upper wing surfaces to clear the cannon breeches and feed motors, and the first sets of 20mm-armed wings were modified on the production line from standard Mk.I (eight gun) wings and fitted to the Hurricane IIc which first entered squadron service in June 1941.

Outclassed
But Still Useful…

Following the Battle of Britain, the Hurricane continued in service, and throughout the Blitz of early 1941 it became the principal single-seat night fighter in Fighter Command. From its earliest service days the Hurricane had proved a pleasant aeroplane to operate at night (unlike the Spitfire with its narrow track undercarriage and sensitive controls), and during the Battle of Britain period Hurricanes regularly flew night patrols to supplement those of the night-fighter Blenheim Ifs. As the *Luftwaffe*'s daylight raids reduced during late September/early October and the nocturnal campaign took precedence, several Hurricane day fighter units converted to night fighter operations.

At this time Britain's night defences were in a poor state with few suitable aircraft able to successfully operate at night, although Blenheim night fighters were gradually being replaced by the more powerful Beaufighters, albeit in small numbers. Airborne Interception (AI) radar was still in its infancy, so unarmed Douglas Havoc bombers were modified and fitted with a 2,700 million candle-power searchlight, developed and built by GEC, in the nose behind a flat transparent screen with power for the light coming from lead-acid batteries in the bomb bay. The aircraft was guided to enemy aircraft by a mix of ground control and its own on-board AI Mk.IV radar. As the Havoc's own armament had been removed, the aircraft was accompanied by a pair of Hurricanes, which, when the target was illuminated, would then attack the enemy bomber and shoot it down.

These composite Turbinlite squadrons, which had been created out of desperation rather than with any real hope of success, achieved little, and, with the rapid development of AI radar and the introduction of the Mosquito night fighter, this wasteful and fruitless experiment was finally abandoned in late 1942.

Twelve Hurricane IIcs were equipped with pilot-operated AI Mk.VI radar in 1942, but after a brief operational deployment with Nos.245 and 247 Squadrons, during which they proved too slow to serve in Europe, they were sent to India in May 1943 and served with No.76 Squadron in the defence of Calcutta until their withdrawal at the end of December.

Photographed over South Wales in 1941, these Hurricane IIbs from No.79 Squadron were based at RAF Fairwood Common, now Swansea airport. The three aircraft, purportedly piloted by the CO, Sqn Ldr G. D. Haysom (leading in Z3745 'NV-B') and his flight commanders, Flt Lt R. P. 'Bee' Beamont (nearest in Z2633 'NV-M'), and Flt Lt L. T. Bryant-Fenn (in Z3156 'NV-F') appear to be in Dark Earth/Dark Green Temperate Land Scheme upper surfaces. Despite being from the same production batch, all three exhibit intriguingly minor differences in the actual demarcations of the camouflage pattern.
Newark Air Museum

As well as Night Fighting duties, the Hurricane became increasingly used as an intruder. Following the Battle of Britain, it had been decided to take the fight back to the enemy and raids were launched into occupied Europe, known as 'Circuses' – small bombing raids against 'fringe' targets escorted by fighters designed to bring the *Luftwaffe* to battle, or 'Rhubarbs' – small scale raids by pairs of fighters or fighter-bombers against targets of opportunity. Neither were particularly effective and RAF losses rose for little tangible return, other than the feeling that the RAF was striking back.

By the beginning of 1941, it was becoming apparent that the development potential of the basic Hurricane design would soon be overshadowed by the new-technology generation of metal monocoque fuselage, thin-wing fighters then being developed.

Although the Hurricane began the year still employed as a day fighter, its days in that role, at least in the European Theatre of Operations (ETO) were numbered, and by the end of the year with the appearance of the Mk.IIb and Mk.IIc, the Hurricane went over entirely to the night, and day, intruder role. Many of the Hurricane night fighter squadrons were now regularly operating over *Luftwaffe* bomber bases in northern France and the Low Countries and during the last six months of 1941 they claimed the destruction of over fifty enemy aircraft, sixteen coastal vessels, 105 road vehicles and seventeen locomotives. Differing little from their daytime counterparts, other than in their black RDM2 Special Night camouflage, these specialist night intruder Hurricanes were simply fitted with anti-glare shields between the exhaust manifolds and the pilot's windscreen.

The Hurricane's ability to carry under-wing loads was also exploited. Using the attachment points for 44-gallon long-range auxiliary fuel tanks, as used on the ferry flights to the Middle East, Hurricane I,

A pair of Hurricane IIas from No.71 (Eagle) Squadron overfly a squadron mate. The Squadron formed in September 1940, manned by American volunteers who had come to the UK to fight with the RAF, hence the term, 'Eagle', a reference to the bald eagle indigenous to that country. Thought to have been photographed whilst working up at RAF Kirton-in-Lindsey in the spring of 1941, the aircraft in the foreground appears to have had an in-service repaint at some point, judging by the slightly non-standard Dark Earth/Dark Green colour demarcation lines, and sports the Night (black) port underwing. The Squadron moved to RAF Martlesham Heath in Suffolk in April 1941 and commenced operations over Europe during May.
Newark Air Museum

P2989, was tested at Boscombe Down, Wiltshire, in April 1941, to carry a 250lb bomb under each wing using a standard Type 2 universal carrier covered by a canoe-shaped streamlined fairing. The trials were successful and production of the twelve-gun Hurricane IIb was thus cleared to include the optional underwing bomb rack fitting. Further trials, this time with a Merlin XX-engined Hurricane IIb, were undertaken which allowed for the carriage of a pair of 500lb bombs (although due to the weight penalty this didn't become a standard operational load), and saw the type move in to the daylight ground-attack role with several squadrons. Despite ostensibly possessing a twelve-gun battery, two of the Mk.IIb's guns had to be removed (one from each four-gun battery) as each bomb rack obscured one of the middle underwing cartridge case ejector chutes.

Operations with bomb-carrying Hurricane IIbs began in the late autumn of 1941 – presumably cannon-armed Mk.IIcs were deemed destructive enough, although the weight of four cannon and two 250lb bombs would have been even more detrimental to the Hurricanes' performance in the skies over Europe

Eight Hurricane squadrons took part in Operation *Jubilee* (the combined forces attack on Dieppe on 19 August 1942), six with bomb-carrying Mk.IIbs, while two, Nos.3 and 87 Squadrons employed cannon-armed Mk.IIcs. *Jubilee* turned out to be the last major operation involving Hurricanes in any numbers in the Northern ETO. Bomb-carrying Hurricane IIbs continued to operate over Europe on nuisance and intruder sorties until late 1942/early 1943 when they were either replaced by the Hawker Typhoon, or re-equipped with the last Mark of Hurricane, the Mk.IV, armed with 40mm cannon and/or rockets for a relatively brief period during the summer and autumn of 1943 before they too were re-equipped with Typhoons, in time for D-Day.

The Vickers Class 'S' 40mm gun was developed in the late 1930s as an aircraft weapon, albeit intended for bomber defence and tested as such in a turret fitted to a modified Vickers Wellington II. In the event it wasn't adopted for bombers, however, once trials at Boscombe Down in September 1941 with Hurricane IIb Z2326 proved successful, it was adopted as an airborne anti-tank gun with special armour-piercing ammunition. Rolls-Royce also developed a 40mm gun, but it never served as an airborne weapon.

Top: Hurricane IIc, BD867 'QO-Y' of No.3 Squadron. Having operated Hurricane Is in the north of Scotland, this unit moved to southern England in April 1941 and re-equipped with Hurricane IIbs and IIcs upon arrival. Initially operating as a night-fighter unit, including a period in which they co-operated (albeit somewhat unsuccessfully) with Turbinlite-equipped Havocs, greater success was forthcoming once they commenced intruder patrols across the English Channel. BD867, which was finished in the then new, (unit applied) Day Fighter Scheme of 'mixed grey' and Dark Green upper surfaces with Medium Sea Grey under surfaces, was fitted with anti-glare shields for nocturnal operations, and served only with No.3 Squadron, meeting its end on 19th August 1942 when it was hit by flak during a ground-attack operation near Dieppe. *Newark Air Museum*

Above: Hurricane IIc, Z3899 'JX-W' of No.1 Squadron which was lost following a mid-air collision over the Isle of Wight with Hurricane BD940 on 22nd November 1941. Another ex-night fighter unit that had moved over to intruder patrol, No.1's aircraft were also repainted in the Day Fighter Scheme introduced in August 1941 which consisted of 'mixed grey' and Dark Green upper surfaces with Medium Sea Grey under surfaces. Nose art is comparatively rare on RAF day fighters but Z3899 features a Native American Indian's head in full war bonnet on the nose – was this an American pilot's aircraft perhaps? Also of interest is the manner in which the rear fuselage Sky band has been truncated around the serial number.
Newark Air Museum

Above: Hurricane I, T9519 'HB-L', No.239 Squadron, *circa* January-May 1942. This Hurricane, built by the Canadian Car & Foundry Corporation, was one of twenty (T9519-T9538) delivered to the UK in late 1940 and January 1941. Having served with Nos.312, 315 and 303 Squadrons prior to arriving with No.239 Squadron, T9519 operated alongside other Hurricanes and Tomahawks until both types were replaced by Mustang Is from May 1942. Thereafter it served with several training units, the last being No.1665 Heavy Conversion Unit. It was SOC in October 1944.

Below left: An unidentified Hurricane coded 'PA-A', belonging to No.55 OTU *circa* 1941. This aircraft appears to be finished in a somewhat well-worn scheme, probably of Dark Earth and Dark Green, with white code letters outlined (unusually) in black. Although the yellow outer ring of the fuselage roundel appears as a dark colour, it is because orthochromatic film that has been used, which turned yellow to a dark colour and also changed the hue of others once the film was exposed. The apparent colour variations on the roundel's outer ring was simply a trick of the light as it played across the fuselage contours. *M Derry collection*

Below: Another Canadian Car & Foundry Corporation-built Hurricane I, AG162 'EH-W', also of No.55 OTU – the unit used several identification code combinations. One of a batch of 300 Hurricanes (AF945-AG344) built in Canada as Mk.Is which were subsequently redesignated as Mk.Xs, AG162 served with No.59 OTU, and was allocated to No.188 Squadron which was briefly equipped with the Hurricane I from September to December 1942 while awaiting Typhoon Ibs. AG162 was transferred to No.55 OTU, a large fighter-training unit which in terms of Hurricanes alone reached a peak of seventy-six aircraft in August 1943. AG162 was SOC in December 1944. *Via PH Butler*

The first squadron to be equipped with Hurricanes fitted with two Vickers 40mm guns, mounted one beneath each wing in conformal fairings, was No.6 Squadron, in the Western Desert in June 1942 where they achieved considerable success, although they also suffered heavy losses, mainly to ground fire. The designation applied to these 40mm gun-armed Hurricanes was Mk.IId – basically a modified 1,280hp Merlin XXII-powered Hurricane IIc with the 20mm cannon removed. A pair of wing-mounted .303 inch mgs were installed – used primarily for ranging and sighting purposes, but also to keep the heads of enemy gunners down – and shackles fitted to take the 40mm gun packs. At least three UK-based squadrons operated the type, No.184 being the first, forming at Colerne, Wiltshire, in December 1942, and Nos.137 and 164 Squadrons, which were only partially equipped with the sub-type in 1943, pending receipt of the 'multi-role' Hurricane IV.

Number 20 squadron, based in the Far East, re-equipped with the Mk.IV in May 1943, equipped with 40mm cannon firing high explosive (HE) ammunition against road and river transports. Tests (undertaken in the Far East) showed a high level of accuracy for the weapon, with an average of 25% of shots fired at tanks striking the target. Attacks with HE were twice as accurate as with Armour Piercing (AP) rounds, possibly because the ballistics were a closer match to the .303 inch mgs used for sighting (the HE shell was lighter and was fired at a higher velocity). By comparison, the practice strike rate of the 60lb rocket projectiles (RP) was only 5% against tank-sized targets.

A new universal wing was developed for the Hurricane, which had the ability to take various loads such as the 40mm gun, up to two 500lb bombs, Smoke Laying Canisters and Rocket Projectiles. The fitting of a more powerful 1,620hp Merlin 24 or 27 engine, and an additional 350lb of armour plate,

Top: An unidentified Hurricane IIc night fighter, possibly coded 'DX-Y' or 'DX-X' of No.245 Squadron whilst the unit was involved in intruder operations over Europe in the spring and early summer of 1942. This image illustrates the rather rough, soot-like effect of the RDM2a Special Night finish (which was prone to rapidly wearing off) that was applied to night fighters and bombers of this period. *Tony O'Toole collection*

Above: A line up of Canadian-built Hurricane XIIs with twelve-gun armament, Hamilton Standard propellers and Packard-built Merlin engines. The aircraft in the foreground carries RCAF serial 5470, with an individual aircraft letter 'L' on the rear fuselage. It probably belonged to one of the RCAF home defence units equipped with the type as indeed were many Canadian-based OTUs. *Tony O'Toole collection*

Below: Canadian Hurricanes of No.135 Squadron, RCAF, at Patricia Bay, BC, in 1943. Following the attack on Pearl Harbor, Western Canada was thrust into a war zone from mid-1942 as Japanese forces attacked the Aleutian Islands. As a measure of defence No.135 'Bulldog' Squadron was formed on 15th June 1942 with Canadian-built Hurricane XIIs which were retained until mid-1944 and, for a brief period (July to October 1942) was allocated the code 'XP'. Although these Hurricanes all seem to have a uniform appearance, a closer look at 'O' and 'E' reveals that the style of national markings and codes had begun to change. *Via PH Butler*

Right: Hurricane IIb, Z5159 'GV-33' of No.134 Squadron, No.151 Wing RAF, photographed while operating from Vaenga near Murmansk, Russia during October 1941. Finished with Dark Earth/Dark Green upper surfaces and Sky Blue undersides, all the Wing's aircraft were fitted with Vokes tropical filters and it is thought they were originally intended for shipment to Malta. The unit's coding system was unique in that only the letter 'G' represented the squadron, in this instance No.134 squadron, the second was the aircraft's individual aircraft letter, while the numeral '33' was applied by the Soviet's after the aircraft had arrived. Z5159, amongst other surviving RAF Hurricanes, was left in Russia on 28th October 1941 when No.151 Wing returned to the UK. *Newark Air Museum*

Below: Hurricane IIcs in the final stages of assembly (in this instance at Hawker's Langley facility), a small fraction of the approximately 14,340 Hurricanes built by all concerned. *T Buttler collection*

resulted in a slightly re-shaped underside radiator housing. Initially designated as the Hurricane IIe, it was quickly changed to become the Hurricane IV.

As early as October 1941, experiments with RPs had been undertaken, and whilst initially not as accurate as the 40mm gun – that is until the pilots learned how to use the rockets and developed their skills and tactics with them – the effect of a salvo of eight 60lb warhead RPs could be devastating.

Following a rocket-firing course at No.1 School of Specialised Low Attack, at Milfield near Berwick during the spring, and a period of working up and honing their tactics in March 1943, during Exercise *Spartan*, by mid-June 1943, No.184 Squadron, then based at Manston, Kent, commenced rocket attacks with their newly delivered Hurricane IVs against enemy shipping in the Flushing Straits. Number 184 Squadron was subsequently joined by Nos.137 and 164 Squadrons, following their own rocket-firing courses, during the summer as part of the *Channel Stop* operations. Other UK-based Mk.IV-equipped units included No.186 (RAF), and Nos.438, 439 and 440 (RCAF) Squadrons, which were operational on this type for varying lengths of time, mainly short periods, before being re-equipped with Typhoons.

Despite the devastation that a salvo of RPs could inflict, the weapon was initially restricted and was not allowed to be taken over enemy territory, so, for the overland 'Rhubards', 40mm cannon were carried whilst the rockets were used for attacks on shipping.

The Hurricane IV was withdrawn from frontline operational use in the ETO in March 1944, but Nos.6 (RAF) and 351 (Yugoslavian) Squadrons continued operating the Mark in Italy and Yugoslavia until after the end of the Second World War, as did No.42 Squadron in Burma.

Bottom left: Built as a Hurricane IV, KZ193, along with KX405, were both selected for conversion to become interim Hurricane V prototypes, albeit temporarily, as both were later returned to their original configuration. The prototype Mk.V proper, NL255, was also built and flown on the grounds that a potential need existed for a 'low attack' Hurricane, however, with ample stocks of both the Mk.IId and Mk.IV in existence the project was terminated. KZ193, seen here as a Mk.V at Boscombe Down, armed with 40mm Vickers cannon and fitted with an armoured radiator, was finished in the standard Day Fighter Scheme with a Sky rear fuselage band, but featured a black spinner on its four-bladed Rotol propeller. It undertook performance and handling trials while at Boscombe Down and, whereas the Mk.IV's flying characteristics were at least felt to be acceptable, those of the Mk.V were considered marginal. *Tony O'Toole collection*

The Middle East
and the Mediterranean

When Operation *Crusader*, the second Allied attempt to relieve Tobruk began on 18 November 1941, air support had increased significantly during the year or so since Operation *Compass*, with RAF and Commonwealth units boasting a strength of about 700 aircraft, including nine squadrons of Hurricanes. However, even before *Crusader*, in fact from June 1941, increasingly heavy Hurricane losses, including recently arrived Mk.IIs had been experienced at the hands of the *Luftwaffe*'s Bf 109Es, dictating their progressive replacement as day fighters by Curtiss Tomahawks and Kittyhawks. By 1942, as the fighting swung back towards El Alamein, the Hurricanes were totally outclassed due to the recent arrival of the Bf 109F – a superb fighter for its day. In addition, from November 1941 in the Libyan desert, the Hurricane had also to contend with yet another formidable opponent, the *Regia Aeronautica*'s new Macchi C.202 *Folgore*, a development of the earlier radial-engined, slow but agile C.200 *Saetta*, which was transformed by the adoption of the powerful, Alfa Romeo-built, Daimler-Benz DB 601Aa inline engine: it too outclassed the Hurricane.

Down, but not out, the Hurricane was to receive a new lease of life as a ground attack aircraft much as it did while operating from Britain at this time. Later deliveries of Hurricane IIbs and Mk.IIcs, despite being fitted with drag-inducing yet essential Vokes tropical carburettor filters, achieved success as fighter-bomber and ground-attack aircraft due to their armament, especially with the four 20mm cannon, and the ability to carry two 250lb bombs on underwing racks. (The first tropicalized Hurricanes – Mk.Is – had appeared following Italy's entry into the war in June 1940. They featured a Vokes air filter in a large 'chin' fairing under the engine cowling and many of these aircraft were sent to North Africa and Malta.)

During the Battle of El Alamein which commenced on the night of 23 October 1942, six squadrons of Hurricanes were available, including No.6 Squadron's recently introduced 40mm cannon-armed Hurricane IIds. The short range of the Bf 109 always reduced its effectiveness during periods of rapid movement, and by now the Germans were outnumbered in the desert, which helped the Hurricanes to take a heavy toll of the retreating German and Italian ground forces, with claims of thirty-nine tanks, 212 lorries and armoured troop-carriers, forty-two artillery guns and 200 various other vehicles plus various fuel and ammunition dumps destroyed, in nearly 850 sorties, for the loss of eleven pilots.

Most of the Hurricane Is sent to the Western Desert were fitted with a Vokes tropical carburettor filter, as illustrated by this otherwise anonymous machine and its enthusiastic pilot. The aircraft appears to have retained its European Theatre, Dark Earth and Dark Green (Temperate Land Scheme) upper surfaces with a centrally-divided Night/White under surface scheme. The film of oil around the nose is worth noting and was the reason why many Hurricane Is were fitted with an oil collector ring around the top quadrant of the nose to avoid oil being blown back across the windscreen.
Tony O'Toole collection

A No.274 Squadron Hurricane I possibly photographed at Amriya, Egypt in late 1940, still with the unit's 'YK' codes in place to the rear of the fuselage roundel and fitted with a Vokes tropical air filter. It appears to have Temperate Land Scheme upper surfaces with Night/White under surfaces and what could be the beginnings of the so-called 'sand & spinach' or 'sand & spaghetti' scheme which was applied to the nose and wing leading edges of many Hurricanes in the Western Desert. The effect was achieved by using either a Mid Stone (or sometimes Aluminium) background with Dark Earth and Dark Green 'mottle' or 'squiggles' over the top. *Newark Air Museum*

A Mid Stone and Dark Earth camouflaged No.73 Squadron Hurricane I, V7544 'TP-S', with the 'sand & spaghetti' scheme applied to its nose and wing leading edges, seen following a forced landing by Flg Off James 'Jas' Storrar at Gazala, Libya, on 2nd February 1941. The damage must have been slight as V7544 was subsequently reported as lost on 15th February 1941, by which time it was operating as part of a No.73 Squadron detachment to El Adem. *Tony O'Toole collection*

Another Hurricane I with the 'sand & spinach/spaghetti' scheme applied to the spinner, nose and wing leading edges. Squadron codes appear to have been deleted making identification of this particular aircraft's parent unit difficult, but it would probably have belonged to Nos.73, 208 or 274 Squadrons all of whom were operating in the area in late 1940/early 1941. *Tony O'Toole collection*

Fall of Greece and Crete

Italy had occupied Albania in the spring of 1939 as a result of Mussolini's desire to assert Italy's interests in the Balkans and from where, on 28 October 1940, Italian forces invaded Greece. The Greek army counter attacked and forced the Italians to retreat and by mid-December the Greeks had repulsed the Italians and occupied nearly a quarter of Albania too – tying down 530,000 Italian troops. In March 1941 a major Italian counterattack failed, so Hitler came to the aid of his ally and invaded Greece through Bulgaria and Yugoslavia on 6 April 1941.

On 17 April, No.33 Squadron, now commanded by Marmaduke Thomas St John 'Pat' Pattle, who was to become the leading Commonwealth 'ace' with up to fifty 'kills' before his death in action on 20 April, was moved from its base at Larissa in eastern Greece to join No.80 Squadron, also flying Hurricanes, at Eleusis in the west. Both squadrons had previously flown Gladiators in Egypt, but over the autumn/winter of 1940 both units had re-equipped with Hurricane Is. Both squadrons were heavily involved in the fighting following the German intervention and losses began to mount, both in air-to-air combat and from the bombing and strafing of their bases. After being forced to retreat to avoid being cut off by the rapid German advance on 20 April, the Greek Army of Epirus surrendered to the Germans and on 23 April 1941 the Greek government sued for peace.

The remnants of both fighter squadrons were withdrawn to Crete on 27 April, with the survivors subsequently flying to Egypt at the end of May following the Battle of Crete which began on 20 May 1941 when Germany launched an airborne invasion named Operation *Mercury*. Replenished in Egypt, both squadrons returned to duty, still with Hurricane Is, with No.33 flying in support of the army in the Western Desert, and present at the Battle of El Alamein, while No.80 was moved to Palestine in support of the Syrian Campaign.

Defence of Malta

The Hurricane also played a crucial role in the defence of Malta. After the Italian declaration of war, the air defence of the island depended on a small number of Sea Gladiators flown by a mix of hastily retrained flying boat pilots and staff officers which together formed the Hal Far Fighter Flight. They had held the line against the *Regia Aeronautica* since the latter's first raid on 11 June 1940.

However, to recap, Hurricanes were already on their way from Britain even before hostilities had broken out in the Mediterranean. As mentioned, a small number of Hurricane Is were retained in Malta in June 1940 after they had flown across France, in two separate groups, on separate dates, before that route was closed to them permanently. Of the eleven that survived the jour-

Hurricane I, V7795, fitted with 44-gallon ferry tanks, en route to Greece on 9th April 1941, flown by Battle of Britain 'ace' Sgt Jack Norwell, accompanied by other Hurricanes and led by a Blenheim. After joining No.80 Squadron in Greece, V7795, still in the Temperate Land Scheme but with 'sand & spinach/spaghetti' applied to the spinner, nose and wing leading edges, was also used by P/O Bill Vale who claimed a number of victories over Axis aircraft while flying this Hurricane. Having survived the debacle of Greece, V7795 was destroyed at Maleme airfield, Crete, circa 18th May 1941.
Tony O'Toole collection

ney as far as Malta, five remained there while the other six flew on to Egypt. To supplement Malta's fighters, twelve Hurricanes had been delivered in August, aboard the elderly aircraft carrier HMS *Argus* during Operation *Hurry*. The further delivery of another dozen Hurricanes, courtesy of *Argus*, during November's tragic Operation *White* has also been related following which just four Hurricanes reached Malta. As a defending fighter it is fair to say that the Hurricane I was superior to those Italian fighters and bombers encountered in 1940, although it was a mistake to underestimate the capabilities of the frequently encountered Fiat CR.42 biplane fighter; when deftly flown it could prove to be a match for the Hurricane.

Enter the *Luftwaffe*
By Christmas 1940, perhaps fourteen Hurricanes (and four Sea Gladiators) remained operational in Malta, although serviceability was an ongoing problem due to a lack of spares: and matters became worse at the beginning of 1941. The Germans, impatient with their Italian allies, began to arrive in Sicily with the deployment of the *Luftwaffe*'s *Fliegerkorps* X from Norway, and in the coming year Hurricane pilots would find themselves encountering a far more determined and better equipped foe.

Six further Hurricanes arrived in Malta from Egypt on 29 January to join a consignment that had arrived in crates as deck cargo on a recently arrived convoy. Slowly the fighter force increased in numbers until a nominal twenty-eight Hurricanes, four Sea Gladiators and three naval Fulmars were available, but not all were fully serviceable. Although this was the largest number of fighters that Malta had yet possessed, a black period for Malta's fighter pilots was about to begin with the arrival of Bf 109E-equipped 7./JG 26 in Sicily, with every one of the *staffel*'s pilots being a highly skilled veteran. Despite never numbering more than a dozen or so aircraft at any one time, with perhaps just eight serviceable, Malta's Hurricane Is were not really a match for these new opponents as many of them were well-worn airframes fitted with bulky Vokes filters which badly affected their top speed. Added to this, many of the pilots were inexperienced, especially compared to the men of 7./JG 26 and deliveries of a better fighter were urgently needed. As Spitfires were not yet available for deployment outside Britain the best that would be offered was the marginally improved Hurricane II with its more powerful Merlin XX engine. Further, as many of Malta's Hurricane Is were equipped with two-speed de Havilland propellers, they were unable to climb fast enough to the higher altitudes favoured by German Ju 88s which meant that they were at a disadvantage as they would always be caught in a slow climb when they met the Bf 109Es which always had the advantage of height. Thus Hurricane losses mounted.

Over the following days twelve more Hurricanes were sent to Malta from Egypt, with experienced pilots, and by the end of March No.261 Squadron was reported to have over thirty Hurricanes on strength, although again, not all were serviceable. These were further reinforced on 3 April 1941 when twelve Hurricane IIbs were flown off HMS *Ark Royal*, which, with their Merlin XXs, enabled them to reach the Bf 109E's altitude faster. All twelve made it to Malta even though one of them had lost its tailwheel and punctured a wing tank whilst taking off from the carrier and another one crashed upon landing.

Hurricane I, P2638 of No.274 Squadron, photographed at either Amriya or Ismailia, Egypt, sporting the unit's distinctive lightning flash on the fuselage side which was applied in lieu of squadron codes. Having previously served with No.3 Squadron RAAF, P2638 was later allocated (in sequence) to Nos.274, 73, 80 and 208 Squadrons. Whilst with the latter unit, this Hurricane was converted into a photo-reconnaissance PR.I and was later shot down by Bf 109Fs while returning from a tactical reconnaissance mission on 24th July 1942.
Tony O'Toole collection

Hurricane I, V7670, arrived in the Middle East in early 1941 only to be captured by the Germans in March 1941 and suffered the ignominy of having its RAF markings replaced by *Luftwaffe* crosses. It is seen here at Gambut in early 1942 after being recaptured and was apparently still in reasonably good condition.
Newark Air Museum

Hurricane I, Z4204 'G' No.74 OTU served with No.80 RAF and No.3 RAAF Squadrons prior to being allocated to No.74 OTU which formed at Aqir, Palestine, on 18th October 1941 with an establishment of fifteen Hurricanes, its role being to train pilots for tactical desert reconnaissance. In July 1942 the unit moved to Rayak, Syria, where Z4204 crashed on take-off on 5th November 1942. *M Derry collection*

Hurricanes on the Takoradi Route. Number '6' is Hurricane IIa DG626, seen at Kano, Nigeria, in September 1941 while en route to Egypt from Takoradi on the Gold Coast of West Africa. In order to complete their journey, Hurricanes were equipped with a pair of 44-gallon long-range tanks and a temporary white finish was applied to the top of the cockpit to deflect heat, and to the top of the fuselage and tailplane to make the aircraft easier to locate should a forced landing occur. DG626 appears to be painted in the Dark Earth and Dark Green, Temperate Land Scheme with light-coloured undersides, probably Sky Blue. Temporary white numerals were applied to all aircraft which flew this route. DG626, originally Mk.I V7061, was converted to a Mk.IIa and re-serialed. Little is known about this aircraft following its arrival in Egypt other than it was SOC on 15th February 1942 as having 'been lost'. *Tony O'Toole collection*

During the next few weeks the tempo eased as many of the *Luftwaffe*'s aircraft were sent to support the fighting in Greece and Yugoslavia and only relatively minor raids and skirmishes, mainly involving the *Regia Aeronautica* took place. More Hurricanes were flown to Malta on 27 April under Operation *Dunlop* when *Ark Royal* returned carrying twenty-four aircraft, all Hurricane IIs, comprised of a mix of the Mk.IIa and Mk.IIb. All but one of the fighters arrived safely.

As No.261 Squadron now possessed over fifty Hurricanes at Hal Far, Takali and Luqa, a new unit, No.185 Squadron was formed from C Flight of No.261 in May 1941and based at Hal Far. It was that month too when most of the *Luftwaffe*'s units started to leave Sicily to take part in Operation *Barbarossa*, the German invasion of Russia, leaving Malta to the *Regia Aeronautica* once more.

A large consignment of about fifty Hurricanes subsequently arrived on the island with Nos.213, 229 and 249 Squadrons being flown from *Ark Royal* and HMS *Furious* during Operation *Splice*. It was originally envisaged that these squadrons would land in Malta, refuel and then fly on to Egypt to meet up with their ground echelons who had gone on before them aboard troopships, but there was a change in plan upon their arrival. Number 249 Squadron was informed that whilst the other two units were to continue to Egypt, they would be staying in Malta. Worse, they would not be able to retain their new Hurricane IIs, it being decided that No.261

Squadron would take them to Egypt instead – as all of the latter's pilots were exhausted and long overdue for a rest. Number 249 Squadron inherited their old aircraft.

After the Germans left Sicily, bomb racks were fitted to some Hurricanes thus creating some of the first 'Hurri-bombers' which then took the offensive to Sicily. A further delivery of Hurricane IIs arrived on 6 June, allowing three fighter squadrons to be fully equipped with Mk.IIs followed by another batch on 14 June, most of which were destined to fly on to the Middle East. Yet more Hurricanes arrived on 27 June, amongst which were some cannon-armed Mk.IIcs – all intended for Malta this time, with yet more arriving on 30 June.

Daytime aerial activity over Malta became comparatively quiet for much of August, although night raids continued which had resulted in the forming of the Malta Night Fighter Unit (MNFU) in July with eight Hurricane IIcs and four Mk.IIbs – its pilots seconded from each of the three day squadrons. (The MNFU was redesignated No.1435 Flight on 4 December 1941.) The MNFU's overall black painted Hurricanes scored their first victory – against an Italian bomber – on the night of 5/6 August.

With the day fighters enjoying a quiet spell, offensive forays continued to be made over Sicily. More Hurricanes arrived for onward delivery to Egypt, of which twenty-two of the new arrivals were retained, the remaining twenty-three flying on to Egypt. On 28 September, No.185 Squadron sent six of its 'Hurri-bombers', each with four 40lb bombs on a pair of improvised bomb racks, to make a series of attacks on Sicilian airfields escorted by other Hurricanes. There they encountered a new opponent: the excellent DB 601-engined Macchi C.202, and one of the Hurricanes was shot down. This potent new fighter significantly outclassed the Hurricane II and allowed the Italians to return to daylight missions over Malta.

On 12 November 1941 a Wing, composed of Nos.242 and 605 Squadrons departed from *Argus* and *Ark Royal*, bound for Malta with thirty-seven Hurricanes. Despite losing three of their number en route, and another which crash-landed upon arrival, Malta now possessed two more experienced fighter squadrons, albeit neither was at full strength.

Because of the successful anti-shipping campaign waged by Malta's air and naval units against Axis shipping supplying Rommel in North Africa, the Germans were compelled to withdraw some *Luftwaffe* units from the Russian front and re-deploy them to Sicily to neutralise Malta once and for all. Thus, at the start of December 1941, large numbers of Ju 88s, accompanied by II./JG 3 and II./JG 53, each equipped with new Bf 109F fighters, and NJG 1, equipped with long-range, night intruder Ju 88Cs, arrived in theatre.

The German onslaught re-commenced in the same vigorous fashion displayed earlier in 1941, with Malta's defences being frequently probed and Hurricanes being

Hurricane IIc, BE482 'EY-Y' of No.80 Squadron, based at Bu Amoud in the Nile Delta, Egypt, early 1943. Following the successful conclusion of Operation *Crusader* and the relief of Tobruk, in which this unit was heavily involved, the squadron returned to the Delta to mainly undertake convoy escort work over the Eastern Mediterranean until early 1944. BE482, finished in the Desert Scheme and post-May 1942 national markings, saw service with Nos.80 and 238 Squadrons before being SOC in March 1944. *Tony O'Toole collection*

Hurricane IIb, HL795 'V', of No.274 Squadron, photographed in mid-1942, possibly at Sidi Heneish in Egypt or one of the many landing grounds (LGs) the unit operated from in that part of the Western Desert during this period of intense activity. The lightning flash was used in various forms by this unit on several aircraft types, not just Hurricanes. HL795, finished in the Desert Scheme with post-May 1942 national markings, was later transferred to No.127 Squadron with whom it was written off following a belly-landing in September 1942. *Via PH Butler*

Hurricane PR.II, 'Kathleen', possibly of No.680 Squadron based at LG.219, Matariyah, Egypt in mid-1943. Number 680 Squadron RAF was formed in February 1943 from 'A' Flight of No.2 Photographic Reconnaissance Unit (PRU), and continued in the PR role operating in North Africa and the Mediterranean. The Hurricanes were finished in overall Bosun Blue with low-visibility red/blue roundels and fitted with cameras in a ventral fairing under the fuselage. The unit converted to Martin Baltimores and de Havilland Mosquitoes in early 1944, deploying to Sicily and Sardinia later in the year. *Tony O'Toole collection*

scrambled many times per day. Number 1435 Flight (nee MNFU) pilots were having trouble too, with night-flying Ju 88s. They were able to fly faster than the Hurricanes and, unsportingly, they constantly weaved and changed height too – making it much harder to intercept them in the dark.

By the beginning of February only twenty-eight Hurricanes remained in Malta – with fuel and ammunition running low and with Messerschmitts roaming over Malta at will attacking targets of opportunity. Hurricanes were in such short supply that No.1435 Flight was forced to fly during the day as well as by night. New pilots arrived via Sunderland flying boat to relieve some of the 'old hands', but what was needed was more and better fighters for them to fly!

Spitfires en route

March commenced with a series of daylight raids by Ju 88s escorted by Bf 109Fs, plus Bf 109 'Jabo' fighter-bombers, with night bombers adding their weight after dark. Although the island's Hurricanes had put up a good show, in reality they had long lost air superiority over Malta with seventeen Hurricanes being lost since the beginning of February alone, but this would change when HMS *Eagle* launched fifteen Spitfires during Operation *Spotter* on 7 March 1942.

The Spitfires were led to Malta by eight Blenheims from Gibraltar, and, with Hurricanes covering their arrival they all landed safely at Takali to re-equip No.249 Squadron. Hurricanes would continue to fly alongside the Spitfires for several more months to come of course, with the Spitfires tackling the escorts and keeping them off the Hurricanes while they went for the bombers – just as intended during the Battle of Britain, but with a lot more success.

More Hurricane IIcs arrived in March, but as more and more Spitfires arrived during the ensuing months, the surviving Hurricanes were used more as fighter-bombers until they were replaced in this role too by Spitfires. The survivors were eventually handed over to the Fleet Air Arm for their Swordfish and Albacore pilots to fly on air-sea rescue patrols and day and night fighter-bomber nuisance raids over Sicily. Later, when Air Vice Marshal Keith Park was appointed AOC Malta, he acquired his own Hurricane II, which he coded 'OK-2'.

Hurricane IIb, BG753 'UP-V' of No.605 Squadron, Hal Far, Malta, *circa* early 1942. During the squadron's transit to the Far East, a small detachment was left at Malta to operate from Hal Far, from January to February 1942, before it was absorbed into No.185 Squadron. Finished in the Temperate Land Scheme with Sky Blue under surfaces, the unit's aircraft briefly carried its codes in white as illustrated by BG753, which later served with No.274 Squadron. By January 1943 the aircraft was with No.134 Squadron operating (initially) in defence of the Suez Canal. Finally, BG753 was transferred to the South African Air Force at the end of June 1944. *Tony O'Toole collection*

An eight-gun Hurricane IIb, BG766, armed with two 250lb bombs (on bomb racks acquired from Beaufighters), seen at Hal Far, Malta, late 1942/early 1943. Somewhat of an enigma, in mid-1943 this aircraft was transferred from RAF to Admiralty charge and operated by the Royal Naval Air Squadron Malta, at Hal Far, which was manned by pilots of the Albacore and Swordfish units operating from Malta along with the survivors from the Fulmar Intruder Flight. Repainted in Extra Dark Sea Grey upper surfaces with Azure Blue (or Sky Blue) under surfaces, the aircraft undertook intruder operations and ASR patrols. BG766 eventually joined No.728 NAS (a fleet requirements unit) at Takali, serving from January to March 1944, following which BG766's record ends abruptly – presumably SOC. *Tony O'Toole collection*

Hurricanes in the Far East

The Hurricane served as a front line fighter in the Far East long after it had been replaced in that capacity in the European, Western Desert and Mediterranean Theatres. The area had, perhaps understandably, been a low priority during the first two years of the conflict from 1939, consequently, when the Japanese declared war there were no modern RAF fighters in the Far East.

Twenty five pilots, many of them veterans of the Battle of Britain, and fifty crated Hurricane IIbs were sent to Singapore to enhance the existing defences. They arrived on 17 January 1942, by which time the Allied fighter squadrons in Singapore, flying Brewster Buffaloes, had been virtually overwhelmed: Imperial Japanese Army Air Force fighter pilots had been totally underestimated; so had its principal fighter, the Nakajima Ki-43 'Oscar'.

Thanks to the efforts of No.151 Maintenance Unit, the Hurricanes were assembled and ready for testing within 48 hours, of which twenty-one were ready for operations within three days. All the Hurricanes were fitted with Rotol constant speed propellers and tropical filters. The weight of the twelve machine guns and the drag caused by the Vokes filter made them slow to climb and unwieldy to manoeuvre at altitude, but they were effective bomber killers. The recently arrived pilots were formed into No.232 Squadron. In addition, No.488 (New Zealand) Squadron, equipped with Buffaloes, converted to Hurricanes. On 18 January, the two squadrons formed the basis of No.226 Group. Number 232 Squadron became operational on 22 January and suffered the first losses, and victories, for the Hurricane in Southeast Asia. Between 27 and 30 January, another forty-eight Hurricanes, all Mk.IIas, arrived aboard the carrier HMS *Indomitable*, from which they flew to airfields near Palembang, Sumatra, in the Netherlands East Indies.

Because of inadequate early warning systems (the first British radar stations became operational only towards the end of February), Japanese air raids were able to destroy some thirty Hurricanes on the ground in Sumatra, many of them in one raid on 7 February. After Japanese landings in Singapore, on 10 February, the remnants of Nos.232 and 488 Squadrons were withdrawn to Palembang. However, Japanese paratroopers began the invasion of Sumatra on 13 February and although Hurricanes attacked six Japanese transport ships on the 14th, they lost seven aircraft in the process. On 18 February, the remaining Allied aircraft and aircrews moved to Java. By this time, only eighteen serviceable Hurricanes remained out of the original force of almost 100.

That same month, twelve Hurricane IIbs were supplied to the Dutch forces on Java,

Just as in the Western Desert, the Hurricane gained a new lease of life as a ground attack aircraft in the Far East after the initial Japanese onslaught had been blunted. Photographed here, an otherwise anonymous RAF Hurricane IIc is being prepared for a bombing mission against Japanese forces in Burma during December 1944. *Tony O'Toole collection*

and the pattern was repeated whereby small numbers of Hurricanes put up a brave fight, but were eventually overwhelmed by superior numbers. However, with their tropical filters removed and with reduced fuel and ammunition loads, they were at least able to stay in a turn with the 'Oscars' they fought. After Java was invaded, some of the New Zealand pilots were evacuated by sea to Australia and one aircraft, V7476, which had not been assembled, was evacuated from Singapore and transferred to the RAAF, becoming the only Hurricane to see service in Australia during the Second World War, albeit on training and other non-combat duties.

When a Japanese carrier task force under the command of Admiral Nagumo made a sortie into the Indian Ocean in April 1942, RAF Hurricanes based on Ceylon saw action against Nagumo's forces during attacks on Colombo on 5 April 1942 and on Trincomalee harbour on 9 April 1942.

On 5 April 1942, Captain Mitsuo Fuchida of the Imperial Japanese Navy, who led the attack on Pearl Harbor, led a strike against Columbo with over fifty Nakajima B5N2 'Kate' torpedo bombers and almost forty Aichi D3A 'Val' dive bombers, escorted by thirty-six Mitsubishi A6M2 'Zero' fighters. They were opposed by thirty-five Hurricane

Above: Hurricane IIb, Z5628 'YB-L', No.17 Squadron seen at Mingaladon, Burma, in early 1942. This aircraft had originally been sent to the Middle East (probably via the Takoradi Route as traces of white paint remain visible on the upper rear fuselage) and then rushed to the Far East in a bid to stem the Japanese advance into Burma. Several of the squadron's aircraft, including Z5628, had their upper wing roundels modified with an additional white ring to make them less like Japan's *hinomaru* 'meatball' marking. This particular aircraft survived the initial Japanese onslaught to be transferred to the Indian Air Force. *Tony O'Toole collection*

Below: Hurricane IIb, BG827 'W', No.273 Squadron, Ratmalana, (now Colombo Airport), Ceylon, (Sri Lanka), September 1943, with Warrant Officer Zayzezeirski standing alongside the wing. Finished in Dark Earth and Dark Green upper surfaces with Sky Blue under surfaces the aircraft features the interim modified national markings with overpainted red areas which were used prior to the introduction of the official SEAC markings. Tragically, Warrant Officer Zayzezeirski was killed on 29th October 1943 after hitting a tree during a dummy attack on Ratmalana while flying BG827. *Tony O'Toole collection*

Is and IIbs of Nos.30 and 258 Squadrons, together with six Fleet Air Arm Fulmars of Nos.803 and 806 Naval Air Squadrons. The Hurricanes went for the attacking bombers, but were engaged by the escorting 'Zeros'. Over twenty Hurricanes were shot down, together with four Fulmars, plus six Swordfish of No.788 NAS that had been surprised in flight by the raid. The RAF claimed eighteen Japanese aircraft destroyed, seven probably destroyed and nine damaged, with one aircraft claimed by a Fulmar and five by anti-aircraft fire – which compared with actual Japanese losses of one A6M2 and six D3As, with a further seven D3As, five B5N2s and three A6M2s damaged.

On 9 April 1942, the Japanese task force sent over ninety B5N2s escorted by some forty A6M2s against Trincomalee port and the nearby China Bay airfield. Sixteen Hurricanes opposed the raid, of which eight were lost and three damaged. They claimed eight Japanese aircraft destroyed, four probably destroyed and at least five damaged, although actual Japanese losses were less, namely, three A6M2s and two B5N2s with a further ten B5N2s damaged.

However, the situation began to change in mid-1942, especially along the Indian border with Burma. Five more Hurricane squadrons became available (three arrived from Britain and two others re-equipped with the type). As their numbers increased, the air battles became a little more even. In August 1943 the Allies created South East Asia Command (SEAC), a new combined command created to take over control and planning of operations against the Japanese in Burma and the Indian Ocean, to which, in November, Admiral Louis Mountbatten was appointed as Commander in Chief.

Ground attack role
Just as it did in the desert, the Hurricane gained a new lease of life as a ground attack aircraft once the Spitfire replaced it in the fighter role, the battles over the Arakan representing the last large-scale use of the Hurricane as a pure day fighter. During 1943, the Hurricane IIc became more common as too did its later variants: the Mk.IId, fitted with 40mm cannon; and the 'universal wing' Mk.IV able to accommodate 40mm cannon, bombs or 60lb RPs as required.

All three variants played a significant part in the fighting until the end of hostilities, the fighting in Burma in 1944/45 being amongst the severest in the South East Asia Theatre, where severe logistical and organisational difficulties, which had crippled earlier efforts to invade Japanese-occupied Burma, *had* to be overcome.

Indian Air Force Hurricanes
Special mention must be made of the Indian Air Force, which played an instrumental role in defeating the Japanese army in Burma. Eight front-line squadrons were eventually equipped with Hurricanes, generally converting from either army co-operation types such as the Lysander or later medium bombers such as the Vultee Vengeance. Having converted to Hurricanes, mainly Mk.IIcs, IAF squadrons were generally tasked with ground attack and tactical reconnaissance missions, providing vital information which was later to change the whole course of the war.

Canadian pilot F/O Rod Lawrence of No.5 Squadron RAF in front of his Hurricane IIc 'M' in India. This aircraft displays Lawrence's personal timber wolf and maple leaf motif which had also been applied to his Curtiss Mohawk IV with which this unit was equipped prior to receiving Hurricanes.
Tony O'Toole collection

A Langley-built Hurricane IIc, LB835, sporting the reduced diameter two-tone blue SEAC markings introduced from the end of 1943, operated by No.4 Squadron IAF, (Indian Air Force - the prefix 'Royal' was approved in March 1945). LB835 was later transferred to No.11 Squadron RAF and was SOC in April 1945.
Newark Air Museum

Hurricane IIc, thought to be LD903 'N' of No.28 Squadron, an army co-operation unit charged with tactical reconnaissance sorties over the Imphal in the spring of 1944, and sporting the white Air Command South East Asia Special Identification Markings across the fin and rudder. LD903 was damaged beyond repair at Jorhat, India on 28 April 1944.
Tony O'Toole collection

40mm cannon-armed Hurricane IId, KX421 'G', photographed whilst serving with No.1 Service Flying Training School (India), (SFTS(I)), based at Ambala, India in early 1945. KX421 came to grief at this location on 14th April 1945 when, having bounced badly on landing, belly-landed and was written off.
Tony O'Toole collection

Two Hurricane IVs belonging to No.42 Squadron, seen at Onbauk, Burma, in early 1945. 'AW-B' is KX802 whilst 'AW-C' at the far side is LF477. Because close support units generally operated from forward-based airstrips effective camouflage remained important, and it was for this reason that many Hurricane units dispensed with the white Air Command South East Asia Special Identification Markings on their upper surfaces and tails which could otherwise have compromised them, either whilst dispersed on the ground or when flying at low altitude. As for their fates; KX802 was reported missing on 20th April 1945, while LF477 remained with the squadron until it was SOC in November 1945.
Tony O'Toole collection

During the Battle for Imphal, in March 1944, their close air support helped the Allied forces to finally break through, the Japanese defeat being turned into a rout with IAF aircraft pursuing them relentlessly through the jungles of Burma. In December 1944, the Arakan offensive began, the objective being to capture the Maya peninsula, Akyab, Ramree Island and to contain the Japanese in the Arakan and prevent them from interfering with the advance of XIV Army. Number 4 Squadron IAF operated in direct support of the land forces, as well as bombing Japanese strongpoints, and during the landing of Indian troops at Kangow the squadron laid a smoke screen on the beach to ensure the troop's safe landing.

IAF Hurricane squadrons were also deployed in support of the second Arakan campaign at the beginning of April, being involved in road blocking sorties and in attacks on specific targets. Number 6 Squadron became a specialist tactical reconnaissance unit supporting XIV Army on this front, and earning the name 'The Eyes of the XIVth Army'. They were also dubbed 'The Arakan Twins' for flying in the standard tactical reconnaissance pairing of leader and weaver. Other operational sorties included providing fighter escort to Dakotas engaged in supply dropping missions in northern Burma. The IAF units were constantly on the move and kept moving from one advanced landing ground to another.

Although IAF Hurricanes were primarily used in the fighter-bomber role in Burma until the end of the war, they occasionally became caught up in air combat too. For example, on 15 February 1944, Flg Off Jagadish Chandra Verma, of No.6 Squadron IAF, shot down a IJAAF Ki-43 'Oscar', claiming the only IAF air-to-air victory of the war.

All at Sea

A Sea Hurricane Ib with arrester hook deployed and flaps down about to land on board a Royal Navy escort carrier (CVE). This image somehow contrives to make the CVE look far bigger than it actually was - it can be taken for granted that at this point, with no view beneath him, the Sea Hurricane pilot would consider the deck to be far too small!
T Buttler collection

Even before the outbreak of war, it had become patently obvious that Fleet Air Arm (FAA) fighter squadrons desperately needed a modern replacement for its antiquated Gloster Sea Gladiators, the dive-bomber-come-fighter Blackburn Skua, and the newly ordered but lacking-in-performance Fairey Fulmar.

Following the Battle of Britain, attacks on Atlantic convoys by Focke-Wulf Fw 200 Condors of *Kampfgeschwader* 40 (KG 40) brought the problem to a head. Not only had the Royal Navy lost two of its already meagre force of seven aircraft carriers, but even with the carriers it still possessed, none could field a fighter with the necessary performance advantage to intercept a Fw 200; even the new Fulmar offered only a minimal speed margin over the Condor. And the Junkers Ju 88 had yet to make its appearance over the shipping lanes!

Neither the RAF nor the FAA had any long-range fighters that could operate from mainland bases, although a suggestion was mooted to operate Bristol Beaufighters from Northern Ireland to cover the Western Approaches, but was rejected, so a shipboard, modern, single-engined fighter had to be found – quickly.

By the mid-November 1940, Fighter Command had begun phasing out its Hurricane Is in favour of the Mk.II, which in effect ostensibly created a pool of surplus Mk.Is that could be relatively easily modified and navalised for use by the FAA. However, while selected airframes were being fully converted to take an arrester hook for conventional service aboard aircraft carriers, in the event, they were not the first of their type to go to sea. In fact the first Hurricanes to operate at sea, in the spring of 1941, belonged to the newly-formed Merchant Ship Fighter Unit (MSFU). Selected merchantmen had a catapult fitted to their bow on which an MSFU Hurricane was perched ready for use. These vessels were termed Catapult Aircraft Merchant ships (CAM ships); their purpose was to provide protection from marauding Fw 200s, Ju 88s, Heinkel He 111s, He 115s and Blohm und Voss Bv 138s. Exceptions to one side, these Hurricanes were in effect 'one-shot' weapons.

Ultimately there were thirty-five CAM ships, all crewed by Merchant Navy personnel, with mainly volunteer RAF pilots supplemented by some FAA pilots. All CAM ship fighters (bar one, aboard SS *Michael E*, the first CAM ship – and the first to be sunk) were placed exclusively under Air Ministry control. In addition to the CAM ships the Royal Navy supplied five Fighter Catapult ships, with FAA pilots, using Hurricane Is and Fulmar Is under the operational control of the Admiralty; two of the five vessels could accommodate up to three fighters.

The first Sea Hurricanes, designated Mk.Ia, were often unofficially referred to as 'Hurricats' and were overhauled and navalised by General Aircraft Ltd who

A pair of Sea Hurricane Ias, Z4867 'LU-Y' and 'LU-P' of the Merchant Ship Fighter Unit (MSFU), on a lighter at Gibraltar. The MSFU was an RAF unit which formed on 5th May 1941 at Speke to provide Hurricanes for CAM ships (catapult aircraft merchantman) to defend against Luftwaffe bomber and reconnaissance aircraft in the days before there were sufficient aircraft carriers. The MSFU used the codes 'KE', 'LU', 'NJ' and 'XS' and was allocated an operational establishment of fifty Sea Hurricanes plus a training establishment of three Hurricane Is and two Sea Hurricanes. The MSFU was eventually disbanded in September 1943. Z4867 was SOC on 29th June 1942. *Crown via PH Butler*

Sea Hurricane Ia, V7504, on a CAM ship's catapult ready to be launched, if or when the necessity arose, in 1941. V7504 had previously served with Nos.303 and 253 Squadrons prior to being allocated to the MSFU, following which it went to No.56 OTU and later still to No.1 Tactical Exercise Unit (TEU), but was damaged beyond repair in March 1944. *Tony O'Toole collection*

Hurricane I, P3206 'NJ-P', although belonging to the MSFU was a pilot trainer and was not equipped for catapulting. When photographed, P3206 had already served with Nos.302, 151, 504 and 303 Squadrons prior to being allocated to the MSFU. Later, this aircraft was transferred to the Second Tactical Air Force Communications Flight and finally SOC on 17th November 1944. *via PH Butler*

received them from the RAF. More than eighty modifications were needed to convert a Hurricane, including: fitting new radios to conform with those used by the FAA; new instrumentation to read in knots rather than miles per hour; the fitting of catapult spools on the underside centre section either side of the radiator housing and either side of the rear fuselage; hoisting shackles to enable the aircraft to be lifted onto a catapult; and underside eye-bolts to lash the aircraft down in case of high wind or bad weather. Most of these aircraft had seen better days, prompting personnel at General Aircraft to refer to the airframes uncharitably as 'neglected and badly maintained cast-offs', which if true was hardly surprising, given that many of the aircraft had been flown intensively with frontline RAF units right up to the time of their transfer to the MSFU.

In all some 250 Mk.Ia conversions were made, with sixty or so equipping the RAF's MSFU at Speke with thirty-five Detached Flights and the FAA's No.804 NAS at Sydenham, Northern Ireland, with four Detached Flights. It is thought that the remainder were allocated to FAA training units. The MSFU Hurricanes concentrated primarily on protecting the Atlantic and Russian convoys, while the Navy's were generally responsible for the UK to Gibraltar route.

Once the Hurricane had been launched off the ship's catapult, the pilot had to either fly to the nearest land or bale out over the convoy and hope to be rescued. Ditching the Hurricane was a poor option as the large radiator air scoop under the fuselage centre section tended to dig in to the water on impact, turning the aircraft onto its back, following which it started to sink almost immediately, along with its disorientated pilot.

The first Hurricat 'kill' was an Fw 200C Condor, shot down on 2 August 1941, with CAM ships remaining in use until the MSFU was disbanded in July 1943, their last escorted convoy being SL133 from Gibraltar during which two Fw 200s were shot down by Hurricanes. Although seemingly suicidal in nature, by the time the CAM Ships and Fighter Catapult Ships were withdrawn from service with the advent of adequate numbers of escort carriers and merchant aircraft carriers (MAC ships), only one pilot had actually been killed while six Fw 200s had been shot down. As important was its deterrent value, the mere reporting of a CAM ship in a convoy is known to have deterred aerial attack and 'snoopers' altogether – itself a valuable asset.

The first carrier borne Sea Hurricanes

Being a more involved conversion, fully navalised, arrester-hooked Sea Hurricanes took a little longer to materialise. In addition to the 'Hurricats' modifications, a V-frame arrester hook was fitted which hinged onto strengthened longerons on the lower fuselage. A new rear fuselage underside section was created with recesses for the arrester hook's arms and a cut-out in the ventral strake to house the actual hook when the unit was retracted.

Following operational trials by No.880 NAS with four Hooked Hurricanes, in July 1941, the Sea Hurricane Ib, as the type was designated, began re-quipping FAA squadrons, initially Nos.801, 880 and 885 NASs aboard the fleet carriers HMS *Eagle*, *Furious*, *Argus* and *Illustrious*. The first Sea Hurricane Ib 'kill' occurred on 31 July 1941 when Sea Hurricanes of No.880 NAS, operating from *Furious*, shot down a Dornier Do 18 flying boat. Other units which subsequently flew the Mk.Ib were Nos.800, 802, 804, 813, 882, 883, 891, 895 and 897 NASs as well as many second-line training squadrons. Sea Hurricane Ibs were also embarked in American-built merchant ship conversions known as escort carriers (CVE), one of which, HMS *Avenger*, was chosen to test the Sea Hurricane's deck landing abilities aboard vessels with flight decks some 330 feet *shorter* than that of an armour-decked fleet carrier. The tests proved successful.

Sea Hurricane Ibs served as far afield as the Atlantic, the Indian Ocean and the Arctic where they helped to escort the Russia-bound convoy PQ18 and the returning QP14 in September 1942. This convoy, the first to Russia following the disaster of PQ17, was the first in the PQ series to receive an aircraft carrier as part of its *immediate* escort. During this trip the Sea Hurricane Ibs of Nos.802 and 883 NASs with twelve aircraft aboard HMS *Avenger* helped to break up every attack made by the *Luftwaffe* on the convoy. Of the thirty or so attackers shot down, sev-

Sea Hurricane Ib, AF974, one of 300 Hurricanes (AF945-AG344) built in Canada as Mk.Is and subsequently re-designated as Mk.Xs. AF974 was transferred to Admiralty charge on 19th September 1941 and allocated to No.880 NAS aboard HMS *Indomitable* on 6th December 1941, coded '7D', and remained with this unit until the following August by which time its code had changed to '7P'. AF974 participated in Operation *Pedestal* in mid-August 1942 still aboard *Indomitable*, then later served with Nos.759, 824 and 769 NASs until disposed of post-March 1944. This image, taken aboard *Indomitable* in late 1941, shows AF974 being marshalled onto the carrier's forward lift. In the background, the 'Sea Hurricane' on the extreme left of the trio, is thought to be BD771, coded '7-Z', a modified ex-RAF Hurricane IIb with an arrester hook section attached under the rear fuselage, which might have been the 'upgunned' Sea Hurricane (a reference to the Mk.IIbs twelve-gun armament) that Sub Lt 'Dickie' Cork flew during *Pedestal*.
M Derry collection

Four Sea Hurricane Ibs ranged on the flight deck of an *Illustrious*-class aircraft carrier, (possibly HMS *Victorious*), including '7U' and '7Y'. All are finished in the Temperate Sea Scheme, comprising Extra Dark Sea Grey and Dark Slate Grey upper surfaces with Sky undersides. '7U' in the foreground also sports a Sky rear fuselage band, and has post-May 1942 style fuselage roundels and fin flashes.
Crown via PH Butler

Sea Hurricane XII, JS327, of No.804 NAS based aboard HMS *Dasher* during Operation *Torch* on 8th November 1942. JS327 was involved in the first day's action, providing top cover for Albacore bombers, but became one of six of the unit's aircraft that ran out of fuel after *Dasher* changed location. Their returning pilots, being unable to find her, were forced to crash-land on Algerian beaches. All participating Fleet Air Arm aircraft, including the Sea Hurricanes, had their roundels modified with a white star applied over them for the duration of *Torch* and were thinly outlined in yellow.
Tony O'Toole collection

Opposite, bottom:
HMS *Nairana* with four No.835 NAS Sea Hurricane IIcs ranged forward. Number 835 NAS was unique in that it repainted its Sea Hurricanes with white upper surfaces, similar to the white 'anti-submarine' scheme carried by the ship's Swordfish, with just the top of the engine cowling left in the original Extra Dark Sea Grey and Dark Slate Grey Temperate Sea Scheme colours to act as an anti-glare panel.
M Derry collection

enteen were credited to the Sea Hurricanes.

Perhaps the Sea Hurricane's most famous action was in covering Operation *Pedestal*, a convoy intended to break the siege of Malta in August 1942 which consisted of fourteen fast merchantmen, including the oil tanker *Ohio*, escorted by two battleships, seven cruisers, twenty destroyers and three fleet carriers, with a fourth carrier conveying Spitfires to Malta. The numerically impressive quantity of fighters aboard the three escorting carriers amounted to: forty-four Sea Hurricane Ibs from Nos.800, 801, 880 and 885 NASs with the Fighter Flight of 813F NAS; twenty Fulmar Is of Nos.809 and 884 NASs; and ten Martlet IIs of No.806 NAS. This convoy would face almost constant air attacks along the whole route by over 650 Axis aircraft based along the North African coast, Sardinia, Sicily and the Italian mainland.

Sea Hurricanes operated from all three carriers: aboard *Eagle* (sunk by a U-Boat on 11 August), was No.801 NAS with twelve Sea Hurricanes and 813F Fighter Flight with four; aboard *Indomitable* was No.800 NAS with twelve Sea Hurricanes and No.880 NAS with ten; while *Victorious* carried just six Sea Hurricanes of No.885 NAS. (The fourth carrier, *Furious*, was loaded with thirty-eight Spitfire Vbs which were to be flown to reinforce RAF units in Malta.)

By the time the convoy's heavy escort of carriers and battleships had to turn away, on the night of the 12/13 August, the only carrier left operational was *Victorious* with eight serviceable Sea Hurricanes, ten Fulmars and three Martlets available. That day, during exceptionally heavy fighting, twenty-nine aircraft had been lost or dumped overboard due to damage. With the loss of *Eagle* and with damage inflicted on *Indomitable* preventing her from flying any aircraft, *Victorious* became the only 'free' deck, consequently many of her unserviceable aircraft (she had also been damaged), had to be dumped overboard to make room for aircraft that were forced to divert from their parent carriers as there was nowhere left to land.

Over the two days of action, FAA fighters had accounted for thirty-nine Axis aircraft along with several more claimed as probably destroyed or damaged for the loss of thirteen of their own in action, (excluding those aircraft lost aboard *Eagle* when she sank). Of these, twenty-five victories were scored by Sea Hurricane pilots for the loss of seven in air combat, one of which was shot down by ships' AA 'friendly fire'. The top scoring FAA pilot (a Sea Hurricane pilot) was Lt 'Dickie' Cork, Senior Pilot in No.880 NAS, who claimed six aircraft shot down (one shared with another pilot) to add to his earlier five 'kills' from the Battle of Britain. Nine of the fourteen merchantmen and two of the escorting cruisers were sunk, yet despite all, enough ships managed to reach Malta, including the single most vital vessel of all, the *Ohio* – barely!

A total of 340 Sea Hurricane Ibs were converted, but the Mark was technically obsolete by 1942, and the Admiralty realised that the Mk.Ib's eight .303 inch machine guns was inadequate when dealing with heavily-armoured enemy aircraft such as the Ju 88 and Bv 138 in particular. It is recorded that about a hundred sets of Hurricane IIc 20mm cannon wings were fitted to navalised Merlin III-engined Hurricane I fuselages to become the Sea Hurricane Ic, which was purported to have entered service in January 1942. The weight of the cannon armament with only a 1,030hp Merlin III, even when modified to accept 16lb of boost, and generating over 1,400hp at low altitude, must have severely affected the type's performance, but details of the type's operational use is vague, and how many airframes were actually produced and whether the type really did enter front-line service is open to speculation.

Bigger-engined Sea Hurricanes

In March 1942, the Admiralty requested some Merlin XX-engined Hurricanes to replace its Merlin III-engined Mk.Is. As an interim, pending the delivery of production, fully navalised, Merlin XX-engined Sea Hurricane IIcs, Hooked Hurricane IIbs and Mk.IIcs, literally RAF standard machines fitted with arrester hooks, and Canadian-built Sea Hurricane IIas, (most of which were retained in Canada) were delivered to the FAA, mainly to shore based training units for type familiarisation.

Fully navalised Sea Hurricane IIcs, built from new by Hawker at Langley, were followed quickly by Canadian-built Packard Merlin-engined Sea Hurricanes, designated Mk.X, powered by a Packard Merlin 28 armed with four 20mm cannon, and Mk.XIIs with the Packard Merlin 29 and armed with twelve .303 inch machine guns. Unfortunately, by the time the main production model Sea Hurricane IIc reached front line units the type was rapidly being superseded, initially by Seafires and American Martlets/Wildcats, and later by Hellcats and Corsairs.

However, Sea Hurricane IIs were involved in one final major event, Operation *Torch*, the codename given to the first Anglo-American amphibious assault of the Second World War, the invasion of French North Africa on the coasts of Vichy-controlled Morocco and Algeria, launched on 8 November 1942. The five squadrons of Sea Hurricanes used during 'Torch' were a mix of

Emphasising just how small escort carriers were, these two Sea Hurricanes (each with two cannon removed) have been positioned on outriggers to enable sufficient clearance for the radar and rocket-equipped Swordfish to pass. One of the Sea Hurricane's drawbacks was its non-folding wing which demanded more deck space per machine than did purpose-designed contemporary naval fighters such as the Wildcat and Hellcat. HMS *Avenger*, for instance, which provided aerial cover for convoy PQ18, carried twelve Sea Hurricanes (plus six dismantled reserves) and three Swordfish for anti-submarine purposes. By comparison the RN would operate twenty to twenty-two of the larger Hellcat in vessels of this class once they became available in sufficient numbers and whose wing span reduced from 42ft 10in to 16ft 2in when folded. *Tony O'Toole collection*

Canadian-built machine gun-armed Mk.XIIs and cannon-armed Mk Xs.

As well as escort duties, in which Sea Hurricanes of No.800 NAS from HMS *Biter*, and those of Nos.804 and 891 NASs from HMS *Dasher*, claimed five French *Armée de l'Air* Dewoitine D.520s from GC III/3 which had attacked Fairey Albacores from *Furious*, other Sea Hurricane sorties during *Torch* included tactical reconnaissance missions, strikes against coastal defences and the airfields of Blida and Maison Blanche and combat air patrols. Six Sea Hurricanes were lost as a result of running out of fuel when their carrier, *Dasher*, moved position and they couldn't locate her, although five managed to crash-land on beaches in American held territory.

FAA Sea Hurricanes operated from Royal Navy fleet carriers and escort carriers for almost three years, scoring an impressive kill-to-loss ratio. The last two frontline operational units to use Sea Hurricanes were composite units, Nos.825 and 835 NASs. Number 825 NAS, aboard the escort carrier HMS *Vindex*, which operated mainly on Atlantic and Russian convoys duties, had twelve anti-submarine Swordfish IIIs plus six Sea Hurricane IIcs, five of which had been modified to fire 60lb rockets – two fitted under each wing – a unique improvisation that was the brainchild of *Vindex*'s Commander Flying, Lt Cdr Percy Gick. Number 835 NAS was aboard *Vindex*'s sister ship, HMS *Nairana*, which also operated twelve anti-submarine Swordfish with six distinctive white-painted Sea Hurricane IIcs, primarily on Atlantic escort duty but also on some Gibraltar convoys too. On 25 May 1944, the convoy they were escorting, MKS 49, was located and shadowed by Junkers Ju 290 reconnaissance aircraft. Although they were driven off, albeit undamaged, one Sea Hurricane failed to pull out of a dive killing the pilot. Then on 26 May, shortly after daybreak, a Mk.IIc, piloted by Sub Lieutenant Burgham shot down Ju 290 9K+FK of FAGr 5 over the Bay of Biscay. That afternoon, Sub Lieutenants Mearns and Wallis attacked two more Ju 290s, Mearns shooting down 9V+GK which ditched in the sea, while the other disappeared on fire into cloud and was assumed to have crashed. Both No.825 and 835 replaced their Sea Hurricanes in September 1944 with Grumman Wildcat VIs. Other Sea Hurricanes served in second line units undertaking vital training functions such as deck landing training, fighter pilot training and co-operation with naval anti-aircraft gunners, until 1945, the year in which the Sea Hurricane retired from naval service.

The Hurricane's Last Gasp

By early 1947, most of the RAF's surviving Hurricanes had been sent to maintenance units where the majority of those that still remained appear to have been finally SOC by July 1947. LF363 was an exception. Today it is familiar to many as it serves with the Battle of Britain Memorial Flight and it is this example which is considered by many to have been the last Hurricane in RAF service – certainly, in mid-1950, it was then the RAF's last flying example. First flown on 1st January 1944, this aircraft served with several units until it was SOC in June 1947 at Middle Wallop, but within two months was 'on charge' again at the same location, becoming the personal mount of AVM Sir Stanley Vincent, DFC, AFC, Senior Air Staff Officer, Fighter Command, who incidentally, was the only RFC/RAF pilot to shoot down enemy aircraft in both World Wars. Painted in an overall silver finish, LF363 had its cannon removed, leaving just their stub fairings, and sported a red propeller spinner with RAF Fighter Command's Badge and an AVM's pennant on its fuselage. In this undated photograph, LF363 still has the AVM's rank pennant displayed aft of the cockpit coaming. LF363 underwent two major services in the post-war period, one in 1948/49 and the other in 1955, before it was selected for formal preservation.
Tony O'Toole collection

By mid-1944, the Hurricane had been largely relegated to second-line duties with the RAF in north-western Europe, its use by the Air Delivery Letter Service being just one example of second-line work. Even in the Far East its RAF days were numbered, as it began to be phased out in favour of other types – although No.42 Squadron continued to operate them until 30 June 1945 when it finally received the P-47 Thunderbolt.

Hurricanes also remained in service as fighter-bombers in the Balkans with No.6 Squadron and in the spring of 1944 had moved to Italy, equipped with RP-armed Hurricane IVs. Axis vessels were attacked at sea along the Yugoslavian coast and amid the numerous creeks and harbours of the Dalmatian islands. Squadron detachments were also made to Bastia in Corsica, Araxos near Patras in Greece, Brindisi, and Ancona. Frequently these Hurricanes carried an asymmetric underwing load consisting of a fixed 44-gallon fuel tank, increasing the Hurricane's duration to almost three hours when cruising at 160mph, and four 60lb RPs – a useful combination when employed on armed reconnaissance missions.

Post-war operations

In July 1945, No.6 Squadron moved to Palestine. Still equipped with RP-armed Hurricane IVs, they worked in cooperation with the police, patrolling the Kirkuk to Haifa oil pipeline to prevent terrorist attacks until, in January 1947, their venerable Hurricanes were replaced by Spitfires – making almost certainly the last *operational* Hurricanes in RAF service.

Number 351 Squadron RAF was a Yugoslav-manned unit that formed in July 1944 at Benina in Libya equipped with Hurricane IIcs in the fighter-bomber role. In September the squadron re-equipped with Hurricane IVs and moved to Canne, Italy, where it joined No.281 Wing RAF, which was part of the Balkan Air Force. Like No.6 Squadron, No.351 was involved in supporting the Yugoslav partisans and operated from an advanced air base on the island of Vis and, from February 1945, was also able to make use of an advance landing ground on the Yugoslav coast. On 5 April 1945, the squadron moved to Prkos, Yugoslavia, where it continued to operate until it disbanded on 16 May 1945. It 'reformed' two days later, in conjunction with No.352 Squadron, to form the 1st Fighter Regiment, Yugoslav Air Force.

Post-war service

The very last of approximately 14,500 Hurricanes built was completed at the Hawker facility, Langley, on 15 September 1944. PZ865, a Mk.IIc, was named 'Last of the Many' in a special ceremony and flown by retiring Hawker Chief Test Pilot 'George' Bulman who, it will be recalled, was the first to fly the prototype almost nine years earlier. PZ865 flies still, today it is with the Battle of Britain Memorial Flight, at Coningsby, Lincolnshire, together with another survivor, LF363. The latter, also a Mk.IIc, was first flown on 1 January 1944 and served with at least four frontline squadrons and several OTUs. Prior to its selection for preservation later on, in 1947, LF363 was serving with No.41 Group Communications Flight where, fittingly, it became the personal mount of AVM Sir Stanley Vincent, DFC, AFC, Senior Air Staff Officer, Fighter Command. Sir Stanley being the only RFC/RAF pilot to shoot down enemy aircraft in both World Wars.

Hurricane IIc, LF380 'FI-D' of No.83 OTU, *circa* June 1944. This unit's role between its establishment in August 1943 and disbandment in October 1944 was to train night bomber crews and at its peak had an establishment of forty Wellingtons and a gunnery flight of four Hurricanes, of which LF380 was one. Finished in the then current Day Fighter Scheme with hastily-applied 'invasion stripes', its four 20mm cannon had been previously removed as they were not required for this particular training role. LF380 was SOC in November 1944.
Tony O'Toole collection

At the conclusion of the Second World War, both at home and abroad, it might be presumed that the Hurricane was immediately disposed of as an obsolete relic. However, the type still remained in operational service with No.6 Squadron in the Mediterranean as late as January 1947 and in service in the UK, albeit in secondary roles, throughout 1946. Here, Hurricane IIc, PG469 'Y3-KI', (Y3 being the unit code) is seen at Aldergrove while being operated by No.518 Squadron, a weather reconnaissance unit which continued to operate the type until October 1946. Struck off charge on 12th August 1946, PG469 appears to have been finished in the Day Fighter Scheme with standard post-May 1942 national markings albeit that its starboard wing roundel appears to be missing; whether this had anything to do with the wing-mounted 'strut' is open to conjecture. *Tony O'Toole collection*

After the Japanese had invaded Java in February 1942, some No.488 (New Zealand) Squadron pilots were evacuated by sea to Australia and one of their aircraft, V7476, which had not yet been assembled, was transported from Singapore with them and transferred to the RAAF, becoming the only Hurricane to see service in Australia during the Second World War, albeit on training and other non-combat duties. V7476 was photographed awaiting scrapping at RAAF Point Cook in 1949 or 1950 with a P-40 in the background.
Tony O'Toole collection

Probably held as a potential donor of spare parts, Hurricane IIc, ex-PG499, was one of the RAF's last Hurricanes, having been retained as maintenance airframe 5500M. It was photographed at RAF Hawarden, (now Chester Hawarden Airport), Flintshire, North Wales, in July 1957, just as the base was closing down. Number 48 Maintenance Unit was formed at RAF Hawarden in September 1939 and until July 1957 it stored, maintained and scrapped many thousands of military aircraft there. Painted in an overall silver scheme with post-war national markings, this particular airframe was sent for scrap in the early 1960s after providing parts for the Science Museum's Mk.I, L1592. As late as 1964 its substantial remains, complete with Merlin engine, still lingered amongst a heap of far more modern (jet) fighters.
M Derry collection

Foreign Operators

The Hurricane had a long operational life with the RAF and Fleet Air Arm, serving in all the theatres of war. It was also exported, and in some cases built, by countries outside of Britain and used by their air forces.

Yugoslavia

The first overseas country to receive Hurricanes was Yugoslavia, who bought twelve Hurricane Is in December 1938. The initial order was followed by a second, also for twelve Mk.Is, in early 1940. Production of a planned 100 licence-built Hurricanes was initiated in Yugoslavia in 1939, forty from the PSFAZ Rogozarski plant in Belgrade and sixty from the state-owned Zmaj factory, of which it is known that some were completed. By April 1941, the *Vazduhoplovstvo Vojske Kraljevine Jugoslavije*, (*VVKJ*, Royal Yugoslav Air Force) had perhaps forty-eight Hurricanes available, including some of the licence-built examples, with one purportedly fitted with a Daimler Benz DB 601 engine which was test flown in 1941. Three units are thought to have been equipped with the type: 51 *Escadrila* 2 *Lovacki Puk;* 33 *Escadrila* 4 *Lovacki Puk;* and 34 *Escadrila,* 4 *Lovacki Puk.*

Later in the war, Yugoslavian personnel manned RAF squadrons that operated Hurricanes at various times; No.351 with Mk.IIcs and Mk.IVs from July 1944 to June 1945 in Italy and the Balkans, and No.352 which was briefly equipped with Hurricane IIcs from April to June 1944 before being re-equipped with Spitfires. After the war approximately sixteen surviving Hurricanes were used by the *Jugoslovensko Ratno Vazduhoplovstvo* (*JRV* – Yugoslav Air Force), initially in the 1st Fighter Regiment (1945), then the Reconnaissance Aviation Regiment (1947-1948) and finally the 103rd Reconnaissance Aviation Regiment (1948-1951).

Belgium

Belgium bought twenty Hurricane Is in mid-1939 and a licence to build eighty more, of which only two or three were completed prior to the outbreak of war. At the start of the Second World War, the *Aviation Militaire Belge,* consisted of three combat *Régiments d'Aéronautique/Luchtmacht* (Aviation Regiments) organised into *Groupes* (Squadrons) generally operating a particular aircraft type which were then further split in to *Escadrilles* (Flights). The *2eme Regiment de Chasse* operated fighters such as the Fiat CR.42, Gloster Gladiator, and the Hawker Hurricane, which were based at Schaffen near Diest. Many of the *Aviation Militaire Belge* Hurricanes were destroyed or badly damaged when they were bombed at Schaffen on 10 May 1940 by *Luftwaffe* aircraft on the first day of the German invasion, the survivors playing an understandably limited part in the ensuing debacle. (With regard to the licence-built aircraft the intention had been to arm these aircraft with four 13.2mm machine guns, three of which had been built and two flown with this armament by May 1940; at least twelve others were built by Avions Fairey with eight .303in mgs.)

Although there were two Belgian-manned fighter squadrons operating under the auspices of the RAF, neither unit was equipped with Hurricanes.

Romania

In early 1939 a Romanian military delegation went to the UK to order fifty Hurricane Is, of which only twelve were delivered to the *Fort ele Aeriene Regale ale României* (Royal Romanian Air Force, FARR), in August 1939. Romania, allied itself with Germany when

Russia was invaded in June 1941, but by that time the survivors of the twelve Hurricanes delivered in 1939 had mainly been relegated to the operational training role and local defence duties with 53 *Escadrila*. However, the first FARR aerial victories of the war were achieved by Lt Horia Agarici flying a Hurricane I of 53 *Escadrila* who shot down three Soviet bombers as they attempted to bomb the Romanian fleet.

Finland
Hawker sold twelve Hurricane Is to Finland in 1939 but they arrived too late to see combat in the Winter War with Russia which started in the November and ended in an uneasy armistice in March 1940. When hostilities began again in June 1941 – the so-called Continuation War, these aircraft were past their best and enjoyed limited use and success, Hurricane pilots claiming only five 'kills'. This was partly due to servicing problems created by a lack of spares and partly because they were worn-out from constant use during the interim uneasy peace between the two wars. One further Hurricane, a Mk.IIb, was captured from the Soviets during the Continuation War and flown by the Finns.

Canada
Two Royal Canadian Air Force (RCAF) squadrons were equipped with Hurricanes, No.1 Squadron – which was re-numbered No.401 Squadron RCAF – was operational in the Battle of Britain, and No.402 Squadron, which formed in December 1940. Several Canadian mainland-based home defence units were also equipped with the Hurricane during the Second World War, and following the implementation of the British Commonwealth Air Training Plan, many Canadian-based OTUs were equipped with the type.

Free French
Hurricanes were used by the *Forces Aériennes Françaises Libres*, (Free French Air Force, FAFL) the air arm of the Free French Forces during the Second World War. Initially operating in the Western Desert, between June 1940 and May 1943, attached to RAF fighter squadrons, *Escadrille de Chasse* 1, later named *Groupe de Chasse* 1, eventually achieved Squadron status when it reformed in the UK as No.341 (Free French) Squadron in January 1943, although by that time it was equipped with Spitfires.

Greece
After the Germans overran Greece the Greek government moved to Allied-controlled Egypt. It established expatriate Greek armed forces units, formed out of personnel that had either been evacuated from Greece or fled to the Middle East. Equipped and organised along British lines, and under British command, No.335 (Greek) Squadron was formed on 10 October 1941 at Aqir, Palestine, with personnel provided from a core of Greek pilots who were undergoing training in Iraq, augmented by others who had fled

from Greece. Initially equipped with Hurricane Is while training, the squadron was transferred to El Daba in Egypt where it was declared operational on 12 February 1942. The squadron began operations over the Western Desert, where it operated continuously until late 1942, participating in convoy protection, bomber escort and ground attack duties. Between June and September 1942 it re-equipped with Hurricane IIbs and participated in the Second Battle of El Alamein in October 1942. In December 1943 the squadron re-equipped with Spitfires.

A second Greek-manned fighter squadron, No.336 (Greek) Squadron, was formed on 25 February 1943 at LG 219 in Egypt equipped with the Hurricane IIc. The unit's first combat mission was on 1 March, being generally employed in shipping protection and air defence duties along the Libyan coast, as well as undertaking fighter sweeps. In October 1943, the squadron began to receive Spitfires.

India
The Indian Air Force, (IAF) played an instrumental role in the war in the Far East, especially in Burma, and operated several squadrons of Hurricanes including Nos.1, 2, 3, 4, 6, 7, 9, 10 Squadrons and No.1 Service Flying and Training School, Ambala. IAF Hurricanes were mainly involved in strike, close air support and aerial reconnaissance duties. In recognition of the services rendered by the IAF, King George VI conferred the prefix 'Royal' in 1945 to become the Royal Indian Air Force – until 1950 when India became a republic and the prefix was dropped.

Iran
The Imperial Iranian Air Force (IIAF) was established in 1920 and by the start of the Second World War its aircraft inventory consisted almost entirely of British and German aircraft. In 1939, two Hurricane Is from an order for eighteen were delivered, the remainder of the order not being fulfilled for several years due to the war. Ten Hurricanes were left by No.74 Squadron RAF in May 1943 when that unit moved to Egypt. Eighteen Hurricane IIcs were delivered in 1946, two of which had been rebuilt as two-seat trainers.

Ireland
During the Second World War, Ireland (Eire), a neutral country, made a request to purchase aircraft from Britain which resulted in thirteen obsolete Hawker Hector biplane light

Finland was another early recipient of the Hurricane, receiving twelve, fabric-winged, Merlin III-powered Mk.Is with de Havilland propellers and unarmoured windscreens. They arrived just too late to take part in the Winter War against Soviet Russia, which broke out in November 1939. Delivered in the Temperate Land Scheme of Dark Earth and Dark Green upper surfaces with Night/White/Aluminium under surfaces, pale blue Finnish swastikas on white discs were applied in the same positions as the RAF roundels. *via PH Butler*

Opposite page:
Yugoslavia was the first overseas country to receive Hurricanes, buying twelve fabric-winged Mk.Is powered by the Merlin II engine driving a Watts two-bladed wooden propeller in December 1938. This particular example was photographed undergoing acceptance trials in the UK immediately after being built – the metal strips over the wing root/outer wing panel join are still in primer. It is finished in the standard RAF Temperate Land Scheme with silver under surfaces and Royal Yugoslav Air Force markings on the wings. *T Buttler collection*

Imperial Iranian Air Force (IIAF) two-seat Hurricane trainer '2-31' seen awaiting delivery at Langley in late 1946. The instructor had the benefit of a Hawker Tempest canopy over his seat. *via PH Butler*

Hurricane I, ex-P5178 of No.79 Squadron RAF, now coded '93' of the Irish Air Corps, with the two-colour style roundels used from 1939 to 1954. P5178 was interned in neutral Ireland during the Battle of Britain after a dogfight with He 111s of KG 55 over the Irish Sea on 29 September 1940 and force landed in County Wexford. Seen in 1942, and still with its RAF Temperate Land Scheme camouflage, this was one of three Hurricanes interned after force-landing in Ireland, although later in the war, several more were purchased, giving the Air Corps a relatively modern fighter for home defence. At its peak some twenty Hurricanes were operated by the force, based at Baldonnel. *via PH Butler*

bombers being supplied during 1941. During the war approximately 163 combat aircraft landed accidentally or force-landed in neutral Ireland and were immediately impounded by the Irish authorities, (although they generally turned a 'blind eye' as the aircrew 'escaped' across the border into Northern Ireland). In this way the *an tAerchór* (Air Corps) acquired such disparate types as a Lockheed Hudson, a Fairey Battle, and, initially, three Hurricanes. Later in the war several more Hurricanes were purchased, giving the Air Corps a relatively modern fighter for home defence, which, at its peak, operated about twenty Hurricanes all based at Baldonnel. After the war, the Hurricanes were replaced by de-navalised Supermarine Seafires and a few two-seat Spitfire trainers.

Netherlands

On 1 January 1942, Dutch forces joined the American-British-Australian Command, but at the onset of the Japanese assault, the *Militaire Luchtvaart van het Koninklijk Nederlands-Indisch Leger* (ML-KNIL, the Military Aviation of the Royal Netherlands East Indies Army), was nowhere near its full combat strength. Of the aircraft ordered, only a small number had been delivered, while many of those in service were obsolete types. However, twelve Hurricane IIbs, diverted from RAF stocks, were supplied to the ML-KNIL on Java where, despite putting up a brave fight, they too were eventually overwhelmed by superior Japanese numbers.

New Zealand

In January 1942, No.488 (New Zealand) Squadron exchanged its Buffaloes for Hurricane IIbs. Following the fall of Singapore the squadron's surviving Hurricanes were transferred to New Zealand for home defence service, where some ended their days as airfield decoys.

No.486 (New Zealand) Squadron, the other RNZAF unit within the RAF to be equipped with Hurricanes, also Mk IIbs, was formed in the UK on 3 March 1942. It operated as a night fighter unit, working in conjunction with No.1453 Flight's Turbinlite Havocs before re-equipping with Typhoons in September 1942.

Norway

Although the *Luftforsvaret* (Royal Norwegian Air Force RNoAF) didn't operate Hurricanes itself, following the occupation of their homeland displaced Norwegian personnel formed two fighter squadrons within the RAF, each financed by the exiled Norwegian government. Number 331, formed in July 1941, was equipped with Hurricane Is from July to August, and Mk.IIbs from August to November 1941 when they received Spitfires.

Poland

Poland's first Hurricanes were bought in 1939 but only one from an order for ten was delivered before the German invasion, the remaining nine being allocated to the RAF or diverted to Turkey instead. However, several expatriate Polish pilots flew Hurricanes within the RAF, with the first Polish-manned squadrons forming in Britain in 1940, of which Nos.302 and 303 (Polish) Squadrons took part in the Battle of Britain.

In total, seven Polish-manned squadrons, each named after a Polish city or individual, operated various Marks of Hurricane, namely: No.302 (City of Poznan); No.303 (Kosciuszko); No.306 (City of Torun); No.308 (City of Krakow); No.315 (City of Deblin); No.316 (City of Warsaw); No.317 (City of Wilno)

Portugal

Initially, fifteen Hurricane IIcs were delivered to Portugal in 1943, followed by another fifty later in the year, allowing three *Esquadrillas* (squadrons) to be formed. The type remained in service until early 1952 when they were replaced by Republic F-47 Thunderbolts.

Soviet Russia

When Germany invaded the Soviet Union in June 1941 the Soviets found themselves under threat on several fronts, stretching from Leningrad to Moscow and the oil fields far to the south. Britain's decision to aid the Soviets involved sending supplies by sea to their northern ports. It was decided to deliver a number of Hurricane IIbs with Nos.81 and 134 Squadrons, which formed No.151 Wing RAF, to provide both protection for the port areas as well as training Soviet pilots and ground crews prior to their receiving Hurricanes for use by the *Voyenno-Vozdushnye Sily Rossii* (VVS Military Air Force). Over 3,000 Hurricanes were sent to Russia, although not all were new, some having seen extensive prior service. Russian pilots considered the Hurricane IIa and IIb to be under-armed, consequently modifications were made at workshops and by ground crews at several frontline locations whereby the .303 inch Brownings were replaced by two ShVAK 20mm cannon and two Berezin UB 12.7mm machine guns. It is believed that some 1,200 Hurricanes were re-armed with Soviet weapons. Some Hurricanes were also modified to carry 100kg bombs.

A number of Hurricanes were adapted to become two-seater trainers, of which some were employed on operational sorties and for courier purposes, while others served at the Glider Aviation School in Saratov towing Antonov A-7 and G-11 transport gliders. Other Soviet modifications included a tactical reconnaissance version with a camera in the rear fuselage and experimental fixed and retractable ski-undercarriages.

By the summer of 1942 it became clear that the Hurricane was no match for the Bf 109F, and they were gradually transferred to home defence PVO regiments in the rear. At least one unit, 26 GIAP in Leningrad, used Hurricanes as night fighters until the spring of 1944. In late 1943/early 1944 some 40mm cannon-equipped Hurricane IIds and Mk.IVs were delivered, but evaluation and conversion training showed that these Hurricanes were inferior to Il-2s already in major production, and so were not used for frontline sorties.

According to Russian archives the total number of Hurricanes received by the Soviets was 3,082, which included Marks IIa, IIb, IIc, IId, and IV, plus Canadian-built Marks X, XI, XII and XIIA. Some twenty-nine frontline VVS fighter regiments were equipped with the type, between 1941 and 1943, and operated over Moscow as well as many other major cities and on the Karelian, Leningrad and Kalinin Fronts.

Turkey

By 1940, the *Türk Hava Kuvvetleri* (Turkish Air Force) had some 370 aircraft of all types in its inventory, making it one of the largest air forces in the Balkans and the Middle East. Although Turkey did not enter the Second World War until February 1945, on the side of the Allies, Turkish Armed Forces, on full alert, were prepared for war following the military alliance between neighbouring Bulgaria and the Axis Powers in March 1941, and the occupation of neighbouring Greece by the Axis Powers in April 1941. Within a year, Turkey's borders were surrounded by German forces in the northwest and west, with Italian forces to the southwest.

New weapons were purchased from any available source, with the *Türk Hava Kuvvetleri* receiving large numbers of aircraft including Spitfires, Fw 190A-3s and Hurricane Is and Mk.IIs. Twelve Mk.Is were delivered during the autumn of 1939, some from the diverted Polish delivery, with deliveries of Hurricane IIcs arriving during 1943/44.

Latvia

In 1939 Latvia ordered 30 Hurricane fighters and paid for them. However due to the start of the Second World War in September 1939 Great Britain never transferred those fighters to Latvia.

Finished in the standard RAF Day Fighter Scheme, with RAF markings replaced by Portuguese 'Order of Christ' roundel/crosses and green/red fin flashes. No.624 was operated by *Esquadrilla* MP in the late-1940s. Its codes and serial were white. *via PH Butler*

Soviet Mk.IIb Z5252 was left in Russia by No.151 Wing RAF at the end of October 1941. 'White 01' seems to be armed with eight .303 mgs rather than twelve, although the ejection chutes for the outer guns are visible.
Tony O'Toole collection

Turkish Hurricane IIc HV608, complete with a white crescent and star on an overall red-painted rudder, seen prior to its ferry flight to Turkey from an RAF unit in Egypt. Finished in the Desert Scheme of Dark Earth and Mid Stone upper surfaces with Azure Blue undersides, HV608's transfer was officially completed at the end of March 1943.
Crown via PH Butler

Hurricane Weapons

Left: One of the two Hurricane IVs converted to Mk.V standard, possibly KZ193, (the other being KX405), powered by the uprated 1,645hp Merlin 32 engine driving a four-blade Rotol propeller, obligingly banking over to display the Vickers 40mm cannon fit and Vokes tropical carburettor filter. Also visible in this view are the cartridge case ejection chutes for the two .303 machine guns that were retained in the wing of the Mk.IV (and Mk.V), for range-sighting purposes when firing the cannon. *T Buttler collection*

Below: A Hurricane IIb of No.402 Squadron RCAF at Warmwell in October 1941 being loaded with 250lb general purpose (GP) bombs. These were disappointing weapons, a result of their thick casings and consequent low explosive charge-to-weight ratio of less than 30 per cent, which in turn meant relatively minimal damage by either blast or fragmentation. Although GP bombs continued in use throughout the war, they were supplemented from 1943 by medium capacity (MC) bombs with thinner casings and a charge-to-weight ratio of 51 per cent. A 250lb MC was made available for use on light fighter-bombers (Hurricanes, Spitfires etc) but it was still too small to be truly effective and it is implicit that the 500lb MC bomb, when available, was much more so. When bomb racks were fitted to the Mk.IIb, it was necessary to remove one machine gun per wing, as the bomb rack and fairing blocked the cartridge case ejection chute exit.
M Derry collection

Right: Hurricane IId, HW719, seen while serving with the Specialised Low Attack Instructors School, a training unit for ground-attack pilots, at RAF Milfield, in early 1943. It is armed with a pair of Vickers 'S' 40mm cannon under its wings and despite the streamlined fairings, this weapons fit had a noticeable drag-inducing effect on the aircraft's performance. *Crown via PH Butler*

Above: Close-up detail of a Vickers 'S' 40 mm cannon. This gun first saw action in the Western Desert from June 1942 following the delivery of twenty-seven of them to that theatre where both they and their installations worked almost flawlessly, any problems encountered being due to the ammunition itself, some of which it was discovered had not been filled with propellant! Rolls-Royce also developed a 40mm cannon and an order for 1,000 was placed in early 1942, however, for a number of reasons this gun proved overly sensitive during air firing trials and the order was cancelled despite the completion of 200 sets of components. None of the RR guns entered service in an airborne capacity. Each 'S' gun carried a fifteen-round magazine with an additional round inside the chamber and used armour-piercing (AP) or high-explosive (HE) shells according to need. The AP round proved effective against all tanks other than Tigers, and was even known to be capable of penetrating the long-barrelled 75mm gun of a Panzer IV. The HE round was found to be particularly effective in the Far East against Japanese 'soft' targets and, surprisingly, against Japanese tanks too, albeit the latter's armour was much thinner than that on German tanks. The 40mm AP Mk.I shell weighed 2lb.7oz and could penetrate 50mm of armour at an angle of 30° to normal flight. In 1943 the AP Mk.V shell entered service, it weighed 3lb – increasing penetration values by approximately 10 per cent. Although the Vickers 'S' was undoubtedly successful, and accurate, rocket projectiles were considered to be more useful tactically as, round-for-round, they were more destructive; moreover, once a fighter had released its RPs (or bombs) it could in theory revert to a fighter role without pause, whereas 40mm-equipped Hurricanes remained very vulnerable. *via PH Butler*

Below: A Hurricane IIc equipped with rocket projectiles (RPs) fitted with 60lb high-explosive (HE) heads. Lethal against tanks when a direct hit was scored, even a near miss might immobilise one. These rockets were equally effective against other land targets – and ships too, providing they didn't impact the water first, in which case the 60lb head would be torn from the RP's body without damage to the target vessel. Usually Hurricanes carried four RPs per wing unless an asymmetric load was called for. Clearly seen here is the blast plate fitted to the wing to protect it from rocket blast – the plate being both heavy and the cause of much drag – though later tests would reveal that providing the RPs were carried at least 9 inches clear of the wing no damage would in fact be caused. *T Buttler collection*

Above: From November 1942 Hurricane IIc, Z3092/G was sent to Boscombe Down to undertake rocket projectile (RP) trials which, in this instance was to trial a Mk.II RP installation consisting of a 250lb bomb attached to (at least) four 3in rocket motors bolted together. Only two or three such RPs were constructed and it would seem that none were actually fired. *Crown via PH Butler*

Below: Hurricane IIc, Z2905, photographed at Langley during trials with 90-gallon ferry tanks in 1942. *via PH Butler*

Hawker Hurricane Variants

Hurricane I
The first production version, produced between 1937 and 1939, was powered by a 1,030hp Rolls-Royce Merlin Mk II or III engine, initially driving a wooden two-bladed, Watts fixed-pitch propeller. The first batches were fitted with fabric-covered wings, and were armed with eight .303 inch Browning mgs. Ongoing improvements and developments saw the Mk.I series fitted with metal-covered wings, and armour plate for the protection of the pilot. Initially two-speed, and later constant-speed, de Havilland and Rotol three-bladed metal propellers were retrofitted until they could be introduced on the Hurricane production line. By September 1939, the RAF had about 500 Hurricanes in service which formed the majority of Fighter Command's home-based squadrons.

For use in North Africa, the Mediterranean and the Far East Theatres, a Vokes tropical carburettor dust filter could be fitted.

Hurricane IIa Series 1
Essentially a Hurricane I powered by the improved Merlin XX engine, which was a little longer than the earlier Merlin resulting in a slightly longer nose profile of some 4.5 inches in front of the cockpit, which made the aircraft somewhat more stable due to the forward shift in the centre of gravity. The new engine used 100 octane fuel and a mix of 30 per cent glycol and 70 per cent water, which was not only a safer mix, but allowed the engine to run approximately 70°C cooler, which gave longer engine life and greater reliability. This variant also introduced a new and slightly longer, aerodynamic Rotol propeller spinner which replaced the 'blunt' Rotol spinner. The Mk.IIa first flew on 11 June 1940 and entered squadron service in late September 1940.

Hurricane IIb (Hurricane IIa Series 2)
The Hurricane IIb was fitted with a new wing mounting twelve .303 inch Browning mgs, and both the Mk.IIb and the eight-gun Mk.IIa Series 2, were able to be fitted with racks allowing them to carry two 250lb (or two 500lb) bombs, or two, fixed, 44-gallon long-range tanks, more than doubling the Hurricane's fuel capacity but lowering the top speed to just over 300mph. For use in North Africa, the Mediterranean and the Far East, the Mk.IIb (and Mk.IIa) were tropicalized, fitted with Vokes carburettor dust filters, and a bottle of water behind the cockpit. Pilots were also issued with a desert survival kit. The first Hurricane IIa Series 2s were built in October 1940 with the Mk.IIbs entering service in April 1941.

Hurricane IIc
The Hurricane IIc replaced machine guns with four Hispano II, 20mm cannon (two per wing) and was introduced in June 1941. The new wing included the ability to carry two 44-gallon long-range fuel tanks and two 250lb or 500lb bombs, although the weight increase was significant (hence it was mainly Mk.IIbs that carried bombs). By mid-1941 the performance of the Hurricane was markedly inferior to the latest *Luftwaffe* fighters, and the Hurricane initially moved over to the night fighter role, and then to the ground-attack/intruder role, giving rise to the term 'Hurri-bomber'. The Mark could be tropicalized and fitted with a Vokes carburettor dust filter.

Hurricane IId
The Hurricane IId was a Mk.IIc, with the 20mm cannon removed and shackles fitted to carry a Vickers 40mm cannon with 16rpg (15 in the magazine plus one in the chamber), it was fitted in a gondola-style pod, one under each wing. A single .303 inch mg in each wing was provided, loaded with tracers, for aiming purposes. The first Mk.IId flew in September 1941 and deliveries started in 1942. The weight of the guns and their drag impacted on the aircraft's performance, but once the pilots had mastered the weapons' trajectory, it proved to be a devastating weapon against enemy armour. The first unit to receive the type, No.6 Squadron in the Western Desert, were nicknamed the 'Flying Can Openers'.

Hurricane T.IIc
Two-seat training version of the Mk.IIc. Only two aircraft were built, both for the Persian Air Force.

Hurricane III
Version of the Hurricane II powered by a Packard-built Merlin engine, in case a shortage of British-built engines occurred. In the event a shortage didn't occur and the Mk.III was abandoned.

Hurricane IV
Another wing modification based upon the Mk.II airframe, and the last major change to the Hurricane. A dedicated ground attack variant, initially designated as the Mk.IIe, the changes became extensive enough for it to become the Mk.IV after the first few had been delivered. Fitted with a 'universal wing' designed to carry either two 250lb or two 500lb bombs, or two 40mm Vickers cannon, or two 44-gallon long-range fuel tanks, or eight 60lb RPs – plus two .303 inch Browning mgs fitted

■ HURRICANE IIc GENERAL CHARACTERISTICS

Power Plant:	Rolls-Royce Merlin XX liquid-cooled V-12 engine rated at 1,185hp at 21,000ft
Maximum speed, clean:	340mph at 21,000ft
full load):	325mph at 18,000ft
Operational range, internal tankage:	460 miles at 175mph cruising speed
with 44-gallon tanks:	600 miles at 175mph cruising speed
Service ceiling:	between 34,000ft and 36,000ft (depending upon load carried)
Rate of climb:	2,750ft/min (clean)
Length:	31ft 5in to 32ft 3in to (depending upon type of propeller spinner)
Wingspan:	40ft 0in
Height, trestled for level flight:	13ft 3in to 13ft 11½ in (depending upon type of propeller blades)
Wing area:	258sq ft
Empty weight:	5,700lb
Loaded weight:	8,044lb
Max take-off weight:	8,710lb
Armament:	4 x 20mm Hispano Mk II cannon
Offensive bomb load:	2 x 250lb or 2 x 500lb bombs

for aiming purpose. The Mk.IV was powered by a 1,620hp Merlin 24 or 27 engine, the Merlin 27 having a redesigned oil system better suited to operations in the tropics which was rated at a slightly lower altitude in keeping with the Hurricane's new role as a close-support fighter. The tropicalized version was fitted with Vokes dust filter. The aircraft also featured additional armour for the pilot, engine and radiator which, in the case of the latter, resulted in a slightly deeper radiator fairing.

Hurricane V
The final variant to be produced of which only three were built/converted and never went into production. Powered by a Merlin 32 engine, boosted to give 1,700hp at low level, the Mark was intended as a dedicated ground-attack aircraft for use in Burma. All three prototypes had four-bladed Rotol propellers. Speed was 326mph at 500ft, which was comparable with the Hurricane I, despite being one and a half times as heavier.

Hurricane X
In 1939, the Canadian Car & Foundry was contracted to produce the Hurricane in Canada. After the CC&F produced 166 Merlin III-powered Hurricanes, (in the AG serial range, still designated Mk.I), a further 268 were built with American manufactured Packard Merlin 28 and designated Mk.X. All 434 Hurricane I/Xs were built to British contracts, with twenty-five being taken over by the RCAF, and given Canadian serials. One Mk.X, RCAF serial 1362 (ex-AG310) was experimentally fitted with a fixed ski landing gear. Initially, the aircraft were produced with an eight-gun armament, although most of those powered by the 1,300hp Packard Merlin 28 engine, (equivalent to the British-made Hurricane II), had twelve machine gun *or* four-cannon wings fitted. CC&F had built over 1,400 Hurricanes, when production was terminated in 1943. (See also Sea Hurricane X)

Hurricane XI
A Canadian Car and Foundry-built variant based on the Hurricane IIa, and essentially similar to the later Hurricane X, (i.e. Mk.II-length version), the Mk.XI designation was applied to aircraft fitted with RCAF equipment mainly for the operational training role, of which some 150 were built, although some were mixed in with Mk.IIs on UK contracts. There also appears to have been a further sub-variant, the Hurricane XIB which is believed to have referred to the installation of the Packard-Merlin 29 engine with a different gear reduction ratio.

Hurricane XII
A Canadian Car and Foundry-built variant based upon the Mk IIb, powered by a 1,300hp Packard Merlin 29 and armed with twelve .303 inch machine guns. (See also Sea Hurricane XII)

Hurricane XIIA
A Canadian Car and Foundry-built fighter-bomber variant, powered by a 1,300hp Packard Merlin 29, armed with eight .303 inch machine guns and able to carry bombs.

Sea Hurricane Ia
The Sea Hurricane Ia was the designation applied to ex-RAF Hurricane Is modified by General Aircraft Limited to be carried by CAM ships. Over eighty modifications were needed including new radios to conform to those used by the FAA, and new instrumentation to read in knots rather than mph. Often referred to as 'Hurricats'.

Sea Hurricane Ib
A fully navalised, eight-gun wing Hurricane I equipped with catapult spools and an arrester hook. A total of 340 aircraft were converted.

Sea Hurricane Ic
About a hundred sets of Hurricane IIc 20mm cannon wings are thought to have been fitted to navalised Merlin III-engined Hurricane I fuselages under the designation of Sea Hurricane Ic. Details of the type's operational use is vague, and how many airframes were actually produced and whether the type really did enter front-line service is open to conjecture.

Sea Hurricane IIc
The primary Sea Hurricane variant based upon the 20mm cannon-armed Hurricane IIc, and fitted with an arrester hook and naval radio gear. Fully navalised Sea Hurricane IIcs, built from new by Hawker at Langley with others converted from Hurricane IIcs, were used on fleet and escort carriers. The Sea Hurricane IIc's Merlin XX engine generated 1460hp at 6,250ft and 1435hp at 11,000ft producing a top speed of 320 mph at 13,500ft and 340mph at 22,000ft.

Sea Hurricane X
As well as building land-based Hurricanes in Canada, the Canadian Car & Foundry was also contracted to build Sea Hurricanes, the Sea Hurricane X, essentially being a Packard Merlin-engined Hurricane X, converted to Sea Hurricane standard and armed with four 20mm cannon.

Sea Hurricane XII
Canadian-built Packard Merlin engined Hurricane XII, converted to Sea Hurricane standard and armed with twelve 0.303 inch machine guns.

Sea Hurricane XIIA
Canadian-built Hurricane X converted for operations aboard CAM ships, armed with eight .303 inch mgs.

Hillson F.40 (FH.40)
A full-scale version of the Hills & Son, slip-wing biplane/monoplane design, using ex-RCAF Hurricane I, 321 (RAF serial L1884) returned from Canada for taxiing and flight trials at RAF Sealand during May 1943, and the Aeroplane and Armament Experimental Establishment, Boscombe Down, from September 1943. The upper wing was not released in flight before the programme was terminated due to poor performance.

Photo Reconnaissance Hurricanes
The Service Depot at Heliopolis in Egypt converted several Hurricanes Is for the photo reconnaissance role. Of the first three, converted in January 1941, two carried a pair of F.24 cameras with 8 inch focal length lenses whilst the third carried one vertical and two oblique F.24s with 14 inch focal length lenses mounted in the rear fuselage, close to the trailing edge of the wing. A streamlined fairing was built up over the lenses aft of the radiator housing. A further five Hurricanes were modified in March 1941 whilst two were converted in a similar manner at Malta in April 1941. During October 1941, a batch of six Hurricane IIs were converted to become PR.IIs, while a final batch, thought to be of twelve aircraft, was converted in late 1941. The PR.II was claimed to be capable of achieving nearly 350mph and able to reach almost 38,000ft.

Tactical Reconnaissance (Tac-R) Hurricanes
Some Hurricanes were converted to become tactical reconnaissance aircraft. An additional radio was fitted for liaison with ground forces who were better placed to direct the Hurricane. Some Tac-R Hurricanes had a vertically angled camera fitted in the rear fuselage, the extra weight requiring either one or two machine guns or two cannon to be omitted. Externally these aircraft were only distinguishable by the 'missing' armament.

Prototype K5083, New Types Park, Hendon Air Pageant, Middlesex, 1936 | Constructed to a new specification, F.36/34, (which would later include the installation of eight 0.303 inch machine guns in the wings), the Hurricane prototype, K5083, was put on show in the New Types Park, at the Hendon Air Pageant in 1936. Finished in overall aluminium doped fabric, the metal cowling, forward fuselage and wing centre section panels were polished metal. Red, white and blue roundels, in the bright pre-war shades, were carried in six positions with the serial number K5083 in black on the rear fuselage. A black numeral '1' (New Type number) was applied to the fuselage in front of the roundels. The prototype had a slightly different fuselage spine, cockpit canopy and underwing centre section compared to the early production Hurricanes.

Hurricane I, L1568, 'S'/'73' of B Flight, No.73 Squadron, RAF Sutton Bridge, Lincolnshire, July 1938 | The Hurricane was ordered into production in June 1936, powered by a 1,030hp Rolls-Royce Merlin II engine driving a Watts, two-bladed, wooden fixed-pitch propeller, and fitted with fabric covered wings armed with eight 0.303 inch Browning machine guns. L1568 was amongst the first 600 airframes, delivered between December 1937 and November 1939 and was finished in the then recently introduced Temperate Land Scheme of Dark Earth and Dark Green upper surfaces with Aluminium painted under surfaces. Red, white, blue and yellow roundels were carried above the wings and on the fuselage sides, with red, white and blue on wing undersides, all in the pre-war 'bright' shades. Allocated to No.73 Squadron, L1568 had a roundel blue individual aircraft letter 'S' on the fuselage sides and the numeral '73' on fin, below a white 'Fighter' spearhead.

Hurricane I, L1643, NO-B of No.85 Squadron, RAF Debden, Essex, winter 1938/1939 | As the war clouds loomed, during the so-called Munich Crisis, the bright pre-war red, white, blue and yellow roundels were toned down, by overpainting the yellow areas and increasing the proportions of the red and blue areas to eliminate the white, although in L1643's case the fuselage roundels retained the yellow outer ring. Also at this time, Fighter Command's aircraft under surfaces were repainted in the distinctive Night (black) and White scheme, so that they could be tracked overland by the Observer Corps, in L1643's instance, correctly divided down the fuselage centreline. Serial numbers were retained under the wings at this time, black under the starboard (White) wing and white under the port (Night) wing.

Hurricane I, L1940, OP-R of No.3 Squadron, RAF Kenley, Surrey, early 1939 | By early 1939, the toning down of national markings was well established, and red/blue roundels painted in the new duller 'wartime' shades were becoming the norm. However, the red, white and blue roundels (still in the pre-war 'bright colours') under the Night/White wings were, like the serial numbers, ordered to be removed, but were sometimes retained, as in L1940's case. Fuselage codes were to be applied in Medium Sea Grey.

Hurricane I, L1584, '111' of No.111 Squadron, RAF Northolt, Middlesex, summer 1938
No 111 Squadron was the first frontline RAF Fighter Command unit to receive the Hurricane, the first examples arriving in late December 1937. Finished in the Dark Earth/Dark Green Temperate Land Scheme upper surfaces, the under surfaces of L1584 were painted in an experimental Night and White scheme, whereby just the outer wing panels were painted, with the fuselage and tailplanes left in (painted) Aluminium (see underside plan view). Serial numbers, in black under the starboard and white under the port wing were retained. The squadron applied its number, 111, in white numerals on the fuselage sides at this time, with the top third in the Flight colour, L1584, having red tops for Red Flight. The squadron badge in a standard frame was carried on the fin.

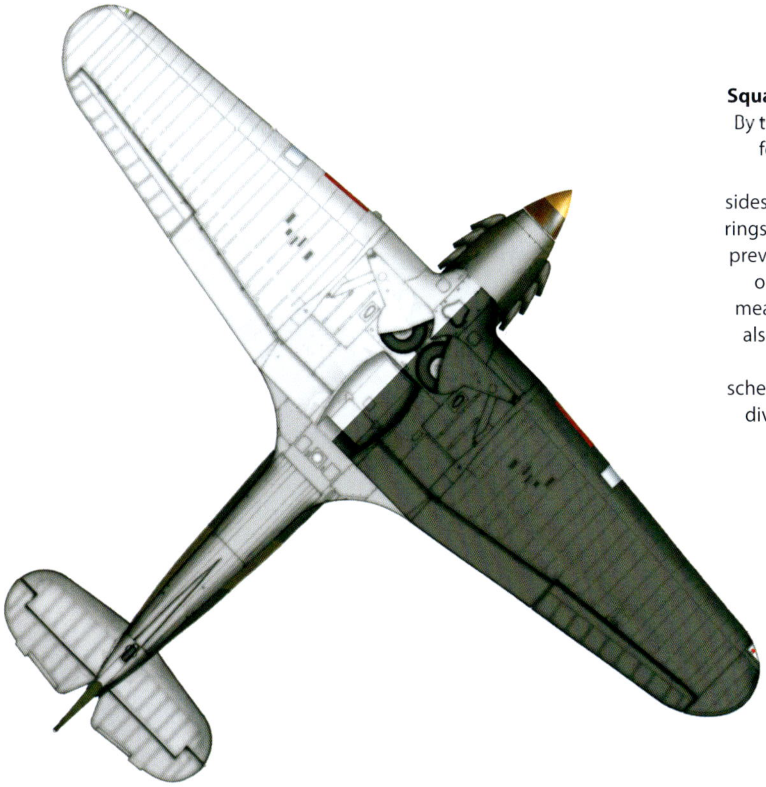

Hurricane I, (serial overpainted), FT-K of No.43 Squadron, RAF Tangmere, West Sussex, September 1939
By the outbreak of war, most Fighter Command Hurricanes featured toned-down national markings, albeit with the pre-war roundels above the wings and on the fuselage sides still showing fresh camouflage where the yellow outer rings had been overpainted. Underwing serial numbers had previously been ordered to be deleted and often the serials on the fuselage sides were overpainted too, as a security measure, as illustrated by FT-K. Additionally, roundels were also ordered to be removed from the wing under surfaces and this aircraft shows a frequently seen under surface scheme variation, whereby the Night/White under surfaces, divided centrally, had (painted) Aluminium applied to the nose, rear fuselage and tailplanes.

Hurricane I, (serial overpainted), LK-O of No.87 Squadron, Lille/Seclin, France, early 1940 | No.87 Squadron was one of the first RAF units to be sent to France upon the outbreak of war in September 1939, and was based at Lille/Seclin, near Calais. By early 1940, the red/blue fuselage roundels had given way to red/white/blue ones, and many of the RAF fighters had adopted French *Armée de l'Air* practices to help with air-to-air recognition, which included red/white/blue rudder stripes, and the re-introduction of underwing red/white/blue roundels for fighters based in France. Hurricanes were also being retrofitted with de Havilland three-blade metal propellers at this time but still retained fabric wings.

Hurricane I, L1814, PO-C of No.46 Squadron, Bardufoss, Norway, May 1940 | Germany's invasion of Norway on 9 April 1940, prompted the British Government to send help, which included No.46 Squadron equipped with Hurricanes. By this time the programme to replace the type's fabric wings with metal ones and the Watts two-blade wooden propellers with three blade metal propellers was gaining momentum, and L1814 was one of the updated airframes sent to Norway. On 1 May 1940, yellow outer rings were ordered to be added to fuselage roundels and red/white/blue stripes painted on the fin – in L1814's case 9 inch wide blue/white stripes were applied with the remainder of the fin in red.

Hurricane I, N2319, VY-P of No.85 Squadron, flown by Sgt Geoffrey 'Sammy' Allard, Seclin, France, May 1940 | N2319 is illustrated as it looked when the 'Blitzkrieg' was under way. Fitted with a DH propeller and a metal 8-gun wing, N2319's under surfaces retained the Night and White scheme, divided down the centreline, with red/white/blue roundels. An Air Ministry signal, X485, was sent on 1 May 1940 ordering that a yellow outer ring should be painted around the fuselage roundels and that red/white/blue stripes be added to the fin, to be implemented with immediate effect. This produced a number of variations; in N2319's instance the yellow outer ring was approximately 4 inches wide and the fin stripes approximately 6 inches wide. It is believed that N2319's propeller spinner was painted yellow, possibly as a Flight marking.

Hurricane I, P3310, DZ-V of No.151 Squadron, RAF Martlesham Heath, Suffolk, June/July 1940 | No.151 Squadron was heavily involved in covering the evacuation from Dunkirk. P3310 was fitted with metal wings and a Rotol constant speed propeller from the outset, and following the introduction of Sky under surfaces in early June 1940, the aircraft had its Night/White undersides overpainted 'in service' in one of the 'Sky' substitutes – possibly a duck egg blue shade – as the official colour Sky was not readily available until later in the summer. No roundels were carried under the wings of home-based fighters at this stage, (they weren't re-introduced until mid-August 1940), but the yellow outer ring, introduced in May 1940, was added to the factory-applied 35 inch diameter red/white/blue fuselage roundel creating an oversized marking that resulted in the yellow being truncated along the lower longeron line and causing the yellow to be painted around the existing squadron code letters. In this instance, 5 or 6 inch wide red/white/blue stripes were applied to the fin.

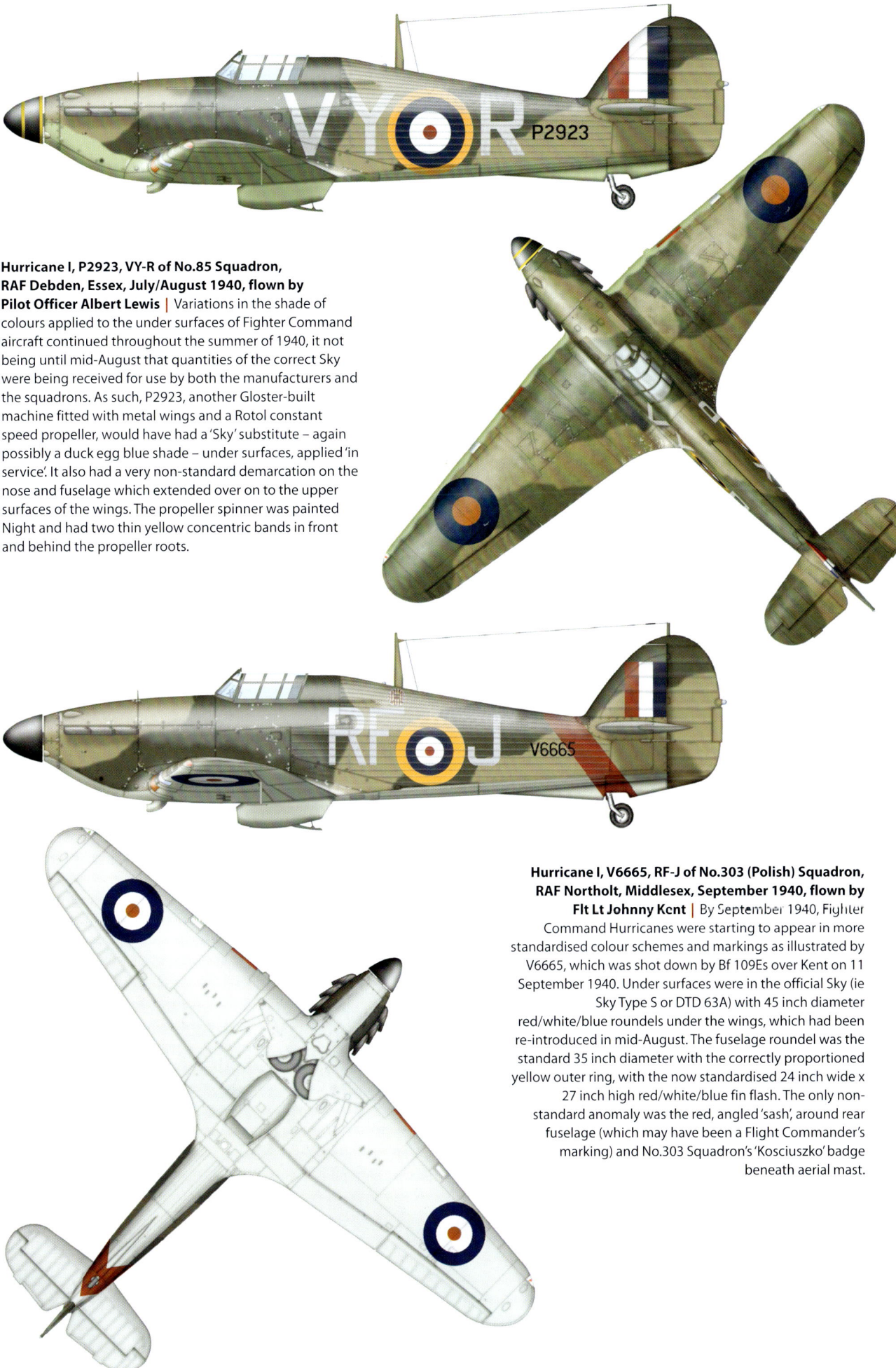

Hurricane I, P2923, VY-R of No.85 Squadron, RAF Debden, Essex, July/August 1940, flown by Pilot Officer Albert Lewis | Variations in the shade of colours applied to the under surfaces of Fighter Command aircraft continued throughout the summer of 1940, it not being until mid-August that quantities of the correct Sky were being received for use by both the manufacturers and the squadrons. As such, P2923, another Gloster-built machine fitted with metal wings and a Rotol constant speed propeller, would have had a 'Sky' substitute – again possibly a duck egg blue shade – under surfaces, applied 'in service'. It also had a very non-standard demarcation on the nose and fuselage which extended over on to the upper surfaces of the wings. The propeller spinner was painted Night and had two thin yellow concentric bands in front and behind the propeller roots.

Hurricane I, V6665, RF-J of No.303 (Polish) Squadron, RAF Northolt, Middlesex, September 1940, flown by Flt Lt Johnny Kent | By September 1940, Fighter Command Hurricanes were starting to appear in more standardised colour schemes and markings as illustrated by V6665, which was shot down by Bf 109Es over Kent on 11 September 1940. Under surfaces were in the official Sky (ie Sky Type S or DTD 63A) with 45 inch diameter red/white/blue roundels under the wings, which had been re-introduced in mid-August. The fuselage roundel was the standard 35 inch diameter with the correctly proportioned yellow outer ring, with the now standardised 24 inch wide x 27 inch high red/white/blue fin flash. The only non-standard anomaly was the red, angled 'sash', around rear fuselage (which may have been a Flight Commander's marking) and No.303 Squadron's 'Kosciuszko' badge beneath aerial mast.

Hurricane I, V7619, XR-F, No.71 (Eagle) Squadron, Kirton-in-Lindsey, Lincolnshire, spring 1941 | In late November 1940, the under surfaces of the port wing of day fighters were once again to be painted black. The roundel on the 'new' black under surfaces was to be given a yellow surrounding ring, which was not supposed to overlap the aileron. To assist further in rapid air-to-air recognition, day fighters were also ordered to have an 18 inch wide Sky band around the fuselage immediately forward of the tailplane and the propeller spinner was to be painted Sky too. However, there is photographic evidence to suggest that initially RAF day fighters had 'light blue' spinners and tailbands, retained well into the summer of 1941, which is thought to have been Sky Blue, as illustrated by V7619.

Hurricane IIb, BD728, FN-T, of No.331 (Norwegian) Squadron, RAF Castletown, Scotland, autumn 1941 | Changes in RAF camouflage policy in August 1941 for UK-based day fighters, saw the introduction of the Day Fighter Scheme, with Ocean Grey replacing Dark Earth, and Medium Sea Grey replacing Sky, which improved camouflage at higher altitudes. However, as with the introduction of Sky under surfaces in mid-1940, due to initial shortages of Ocean Grey, squadrons were asked to mix their own grey shade by mixing seven parts Medium Sea Grey to one part Night, that could be applied in-service in conjunction with the existing Dark Green. Under surfaces were repainted in Medium Sea Grey. BD728, served with No.331 Squadron, the first RAF Norwegian Fighter squadron, when it was repainted, which was presumably when the position of the serial number was moved forward from its usual/standard position.

Hurricane IIc, Z3899, JX-W, No.1 Squadron, RAF Tangmere, West Sussex, November 1941 | At the beginning of 1941, No 1 Squadron was involved in 'Circus' operations, escorting RAF day bombers against targets in northern France and Belgium and 'Rhubarbs' (fighter sweeps), which it continued throughout the year. Z3899 was also finished in the 'Mixed Grey' and Dark Green upper surface scheme, with Medium Sea Grey under surfaces. Remaining part of the Day Fighter Scheme was the retention of the Sky spinner and rear fuselage band, and the introduction of yellow wing leading edges from approximately mid span to the wingtip. What was fairly unique about Z3899 was that it sported nose art, in the form of a Native American Indian's head in full war bonnet under the port exhaust manifolds. This aircraft was lost on 22 November 1941 after colliding with another Hurricane over the Isle of Wight.

Hurricane IIc, BN185, QO-A of No.3 Squadron, RAF Hunsdon, Hertfordshire, September 1942 | During the summer of 1941, No 3 Squadron started night fighter patrols over the London area as well as 'Rhubarb' intruder operations, attacking targets of opportunity in northern France and Belgium and enemy shipping around the coast. At the time it was deemed that black was the best colour for night operations and many of No.3's Hurricanes had their Day Fighter scheme overpainted in a matt black distemper called Special Night RDM2. National markings of reduced diameter and dimensions were introduced in May 1942, with the code letters and serials repainted dull red for aircraft involved in Night Fighting/Intruder missions, as illustrated by BN185.

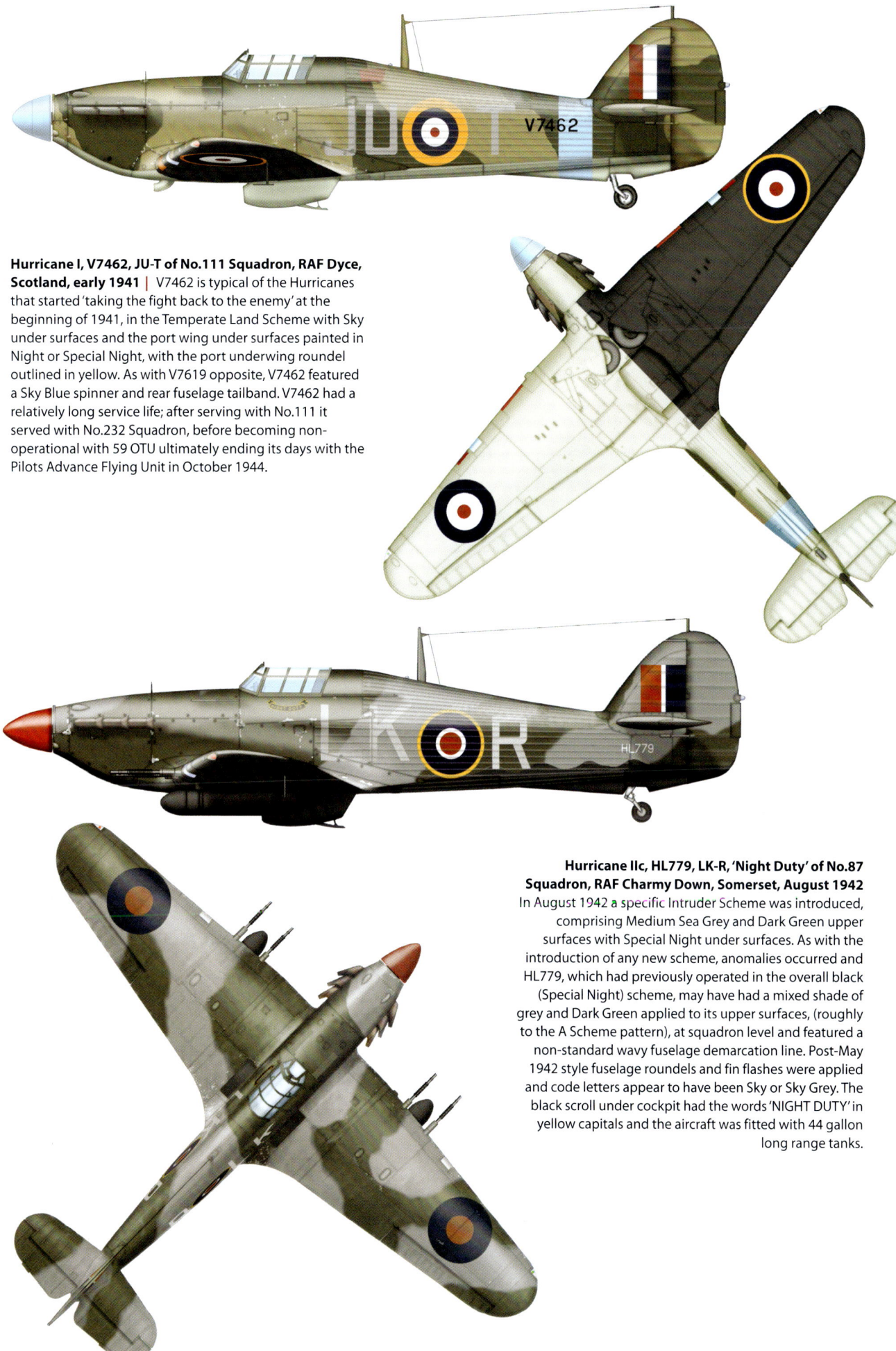

Hurricane I, V7462, JU-T of No.111 Squadron, RAF Dyce, Scotland, early 1941 | V7462 is typical of the Hurricanes that started 'taking the fight back to the enemy' at the beginning of 1941, in the Temperate Land Scheme with Sky under surfaces and the port wing under surfaces painted in Night or Special Night, with the port underwing roundel outlined in yellow. As with V7619 opposite, V7462 featured a Sky Blue spinner and rear fuselage tailband. V7462 had a relatively long service life; after serving with No.111 it served with No.232 Squadron, before becoming non-operational with 59 OTU ultimately ending its days with the Pilots Advance Flying Unit in October 1944.

Hurricane IIc, HL779, LK-R, 'Night Duty' of No.87 Squadron, RAF Charmy Down, Somerset, August 1942 In August 1942 a specific Intruder Scheme was introduced, comprising Medium Sea Grey and Dark Green upper surfaces with Special Night under surfaces. As with the introduction of any new scheme, anomalies occurred and HL779, which had previously operated in the overall black (Special Night) scheme, may have had a mixed shade of grey and Dark Green applied to its upper surfaces, (roughly to the A Scheme pattern), at squadron level and featured a non-standard wavy fuselage demarcation line. Post-May 1942 style fuselage roundels and fin flashes were applied and code letters appear to have been Sky or Sky Grey. The black scroll under cockpit had the words 'NIGHT DUTY' in yellow capitals and the aircraft was fitted with 44 gallon long range tanks.

Hurricane I (trop), P3731 'J' of No.418 Flight, aboard HMS *Argus*, Operation *Hurry*, August 1940 | In late July 1940, a dozen Hurricane Is were formed in to 418 Flight and embarked aboard HMS *Argus* for delivery to Malta where they were desperately needed to reinforce the beleaguered garrison there. Several of the aircraft, including P3731, had their Dark Earth and Dark Green upper surfaces modified with the Dark Green areas overpainted in a 'light brown', (possibly Light Earth or an initial batch of what was later to become Middle Stone), to create a 'desert' scheme. The Night/White under surfaces were retained but a Vokes tropical carburettor filter was fitted. All twelve Hurricanes were successfully delivered to Malta, where they formed part of a new No.261 Squadron.

Hurricane I, V7589, YK-Q, of No.80 Squadron, Yannina, Greece, February 1941 | On 28 October 1940, Italian forces invaded Greece, and No.80 Squadron and its Hurricanes were sent to help defend the country, initially faring well against the *Regia Aeronautica* – before the *Luftwaffe* came to the aid of its Italian ally in April 1941. Finished in the Temperate Land Scheme with Sky under surfaces, V7589 was flown by the legendary Marmaduke Thomas St John 'Pat' Pattle, a Flight Commander with No.80 Squadron at the time, who was to become the leading Commonwealth 'ace' with up to fifty 'kills' before his death in action on 20 April, who downed two Fiat BR 20s flying it on 28 February 1941.

Hurricane IIb, BD789, HA-C, of No.126 Squadron, Ta' Qali, Malta, November 1941 | No.126 Squadron was reformed in June 1941 from a Flight of No.46 Squadron en route for the Middle East and was stationed at Ta' Qali in Malta to provide help with the air defence of the island. Fitted with the twelve machine gun wing and finished in Dark Earth and Dark Green upper surfaces, with Sky Blue under surfaces, BD789 operated from the island until it was lost to enemy action on 8 May 1942.

Hurricane IIb, Z5348, AX-M of No.1 Squadron, South African Air Force, Landing Ground (LG) 92, Egypt, July 1942, flown by Lt Jerks Macle, SAAF | No.1 Squadron SAAF moved to Egypt in April 1941, in support of various allied operations including the First Battle of El Alamein along the northern coast of Egypt in July 1942. On 3 July, the squadron intercepted a formation of Ju 87s escorted by Bf 109s and claimed fourteen Ju 87s and one Bf 109 destroyed for no loss. Z5348, was finished in the Dark Earth and Middle Stone upper surface Desert Scheme, with Sky Blue under surfaces, and featured a cartoon of a man reclining in a chair and a single swastika 'kill' marking under the cockpit. The 4 inch high serial number presentation is interesting.

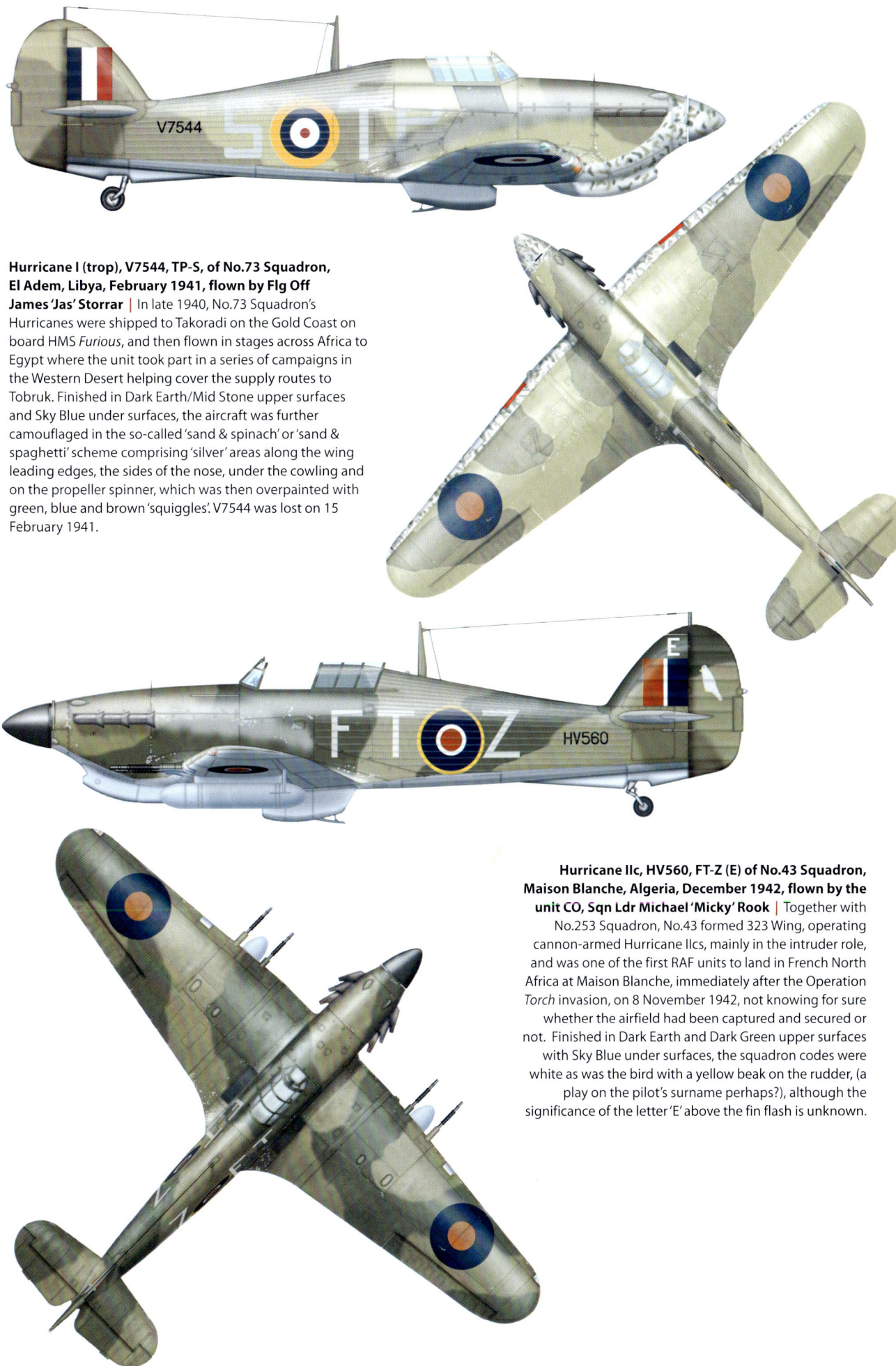

Hurricane I (trop), V7544, TP-S, of No.73 Squadron, El Adem, Libya, February 1941, flown by Flg Off James 'Jas' Storrar | In late 1940, No.73 Squadron's Hurricanes were shipped to Takoradi on the Gold Coast on board HMS *Furious*, and then flown in stages across Africa to Egypt where the unit took part in a series of campaigns in the Western Desert helping cover the supply routes to Tobruk. Finished in Dark Earth/Mid Stone upper surfaces and Sky Blue under surfaces, the aircraft was further camouflaged in the so-called 'sand & spinach' or 'sand & spaghetti' scheme comprising 'silver' areas along the wing leading edges, the sides of the nose, under the cowling and on the propeller spinner, which was then overpainted with green, blue and brown 'squiggles'. V7544 was lost on 15 February 1941.

Hurricane IIc, HV560, FT-Z (E) of No.43 Squadron, Maison Blanche, Algeria, December 1942, flown by the unit CO, Sqn Ldr Michael 'Micky' Rook | Together with No.253 Squadron, No.43 formed 323 Wing, operating cannon-armed Hurricane IIcs, mainly in the intruder role, and was one of the first RAF units to land in French North Africa at Maison Blanche, immediately after the Operation *Torch* invasion, on 8 November 1942, not knowing for sure whether the airfield had been captured and secured or not. Finished in Dark Earth and Dark Green upper surfaces with Sky Blue under surfaces, the squadron codes were white as was the bird with a yellow beak on the rudder, (a play on the pilot's surname perhaps?), although the significance of the letter 'E' above the fin flash is unknown.

Hurricane IIb (trop), BN209, RD-P of No.67 Sqn, Toungoo, Burma, early 1942 | No.67 re-formed at Kallang in March 1941, initially equipped with Brewster Buffaloes, an aircraft that was outclassed in Europe but considered adequate for the Far East. However, the inadequacy of the Buffaloes led to their replacement with Hurricanes in February 1942, but the Japanese offensive could not be halted and by March, the squadron ceased to be effective and re-assembled at Alipore in India, forming part of the defence of Calcutta. Finished in Dark Earth and Dark Green upper surfaces with Sky Blue under surfaces, the fate of BN209 is unsure, but was probably lost to enemy action.

Hurricane IIb (trop), BE208, 'O' of No.232 Squadron, Kallang, Singapore, January 1942, flown by the unit CO, Sqn Ldr R E P Booker, DFC
No.232 arrived at Kallang in the southeastern part of Singapore on 17 January 1942 and was immediately caught up in the Japanese onslaught, often operating against odds of ten to one or worse. Finished in Dark Earth and Dark Green upper surfaces with Sky Blue undersides, BE208, ostensibly the CO's aircraft, carried a rendition of the squadron's badge, a 'Dragonship' (from the Arms of Lerwick, reflecting the unit's association with the Shetland and Western Isles) and a Squadron Leader's pennant under the cockpit. BE208 force-landed at Kallang on 5 February 1942 with Flight Lieutenant E E 'Ricky" Wright at the controls, but its ultimate fate is unknown.

Hurricane IIc (trop), LB830 'R' of No.11 Squadron, Lalmai, India, early 1944, flown by Flt Lt Ronald Erling Rorvik | In August 1943, No.11 exchanged its Blenheim bombers for Hurricane IIcs, becoming a fighter squadron, and moved to Lalmai, India, undertaking escort duties for Vultee Vengeance bombers and supply-dropping Dakotas, although the trend moved more to an offensive ground attack role in support of ground forces during 1944. Finished in Dark Earth and Dark Green upper surfaces with Medium Sea Grey under surfaces and sporting dark blue/pale blue SEAC roundels and fin flashes, LB830 carried a Norwegian flag in a shield under the exhaust manifolds with the words *'Bretoen'* and *'Norge'* written in a Gothic script above and below, Norwegian spellings of Breton and Norway as the pilot's mother was Breton French and his father Norwegian.

Hurricane IV (trop), KX802, AW-B of No.42 Squadron, Onbauk, Burma, early 1945 | No.42 was another Blenheim bomber squadron that was re-equipped with Hurricanes for ground support work, in October 1943, and was still operating the type in early 1943, albeit the dedicated ground attack Mk IV variant, fitted with the universal wing which could carry bombs, 40mm guns or 60lb Rocket Projectiles. Finished in the Temperate Land Scheme with Medium Sea Grey under surfaces which was adopted for fighter aircraft operating in the Far East in the spring of 1944, and dark blue/pale blue 'SEAC' roundels and fin flashes, KX802 'went missing' on 20 April 1945.

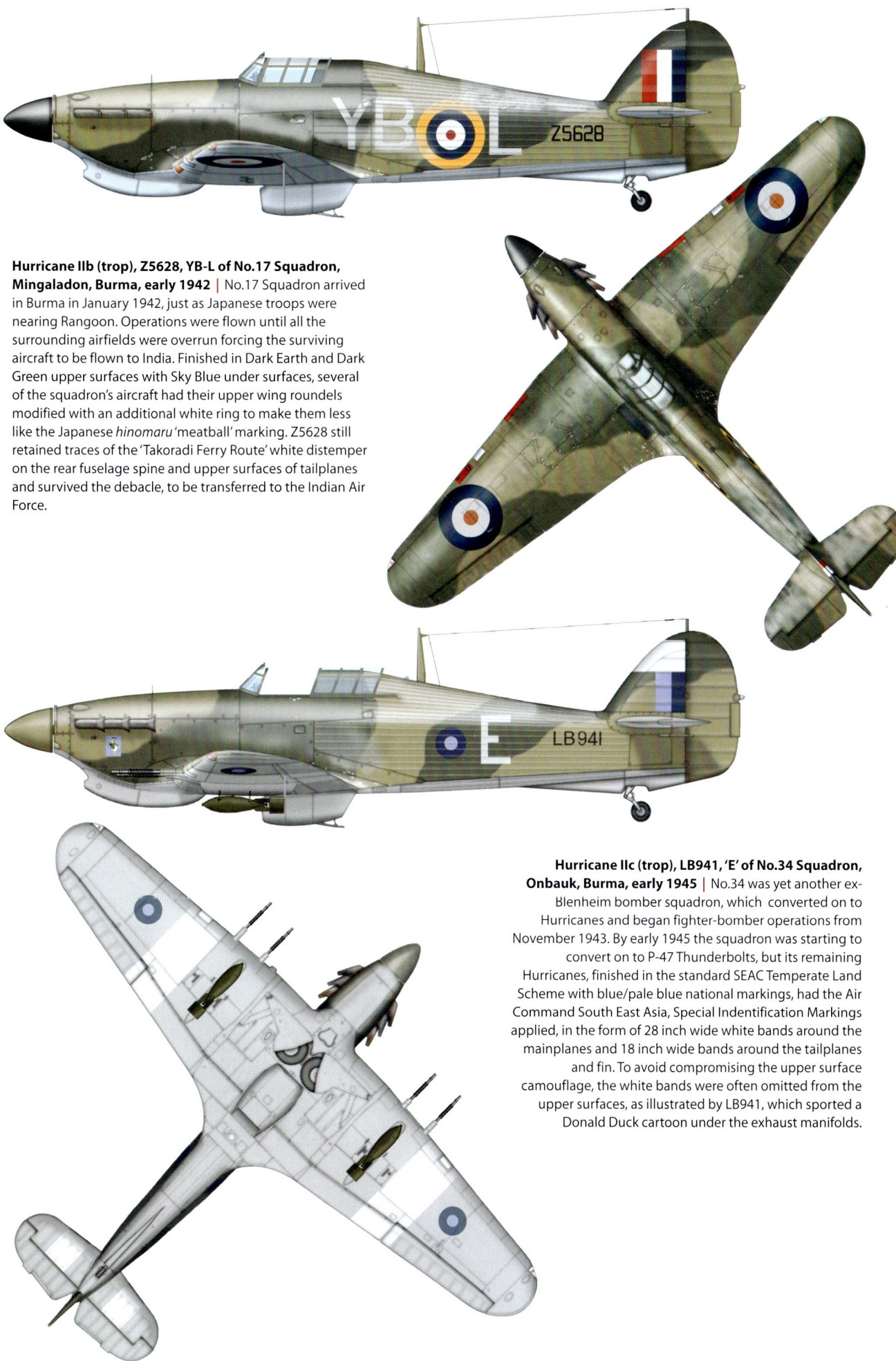

Hurricane IIb (trop), Z5628, YB-L of No.17 Squadron, Mingaladon, Burma, early 1942 | No.17 Squadron arrived in Burma in January 1942, just as Japanese troops were nearing Rangoon. Operations were flown until all the surrounding airfields were overrun forcing the surviving aircraft to be flown to India. Finished in Dark Earth and Dark Green upper surfaces with Sky Blue under surfaces, several of the squadron's aircraft had their upper wing roundels modified with an additional white ring to make them less like the Japanese *hinomaru* 'meatball' marking. Z5628 still retained traces of the 'Takoradi Ferry Route' white distemper on the rear fuselage spine and upper surfaces of tailplanes and survived the debacle, to be transferred to the Indian Air Force.

Hurricane IIc (trop), LB941, 'E' of No.34 Squadron, Onbauk, Burma, early 1945 | No.34 was yet another ex-Blenheim bomber squadron, which converted on to Hurricanes and began fighter-bomber operations from November 1943. By early 1945 the squadron was starting to convert on to P-47 Thunderbolts, but its remaining Hurricanes, finished in the standard SEAC Temperate Land Scheme with blue/pale blue national markings, had the Air Command South East Asia, Special Indentification Markings applied, in the form of 28 inch wide white bands around the mainplanes and 18 inch wide bands around the tailplanes and fin. To avoid compromising the upper surface camouflage, the white bands were often omitted from the upper surfaces, as illustrated by LB941, which sported a Donald Duck cartoon under the exhaust manifolds.

Sea Hurricane Ia, P2921, XS-V of the Merchant Ship Fighter Unit, RCAF Station Dartmouth, Nova Scotia, Canada, mid-1941 | The Merchant Ship Fighter Unit (MSFU) was an RAF unit formed in May 1941 with its headquarters at RAF Speke, near Liverpool. It operated modified Hurricanes from merchant ships fitted with catapults on the bow, referred to as Catapult Aircraft Merchant (CAM) ships. P2921, which had previously served with three RAF frontline squadrons before being modified, was finished in the Temperate Sea Scheme of Extra Dark Sea Grey and Dark Slate Grey upper surfaces, with Sky under surfaces, and is illustrated as it looked whilst temporarily operating from RCAF Station Dartmouth, located on the eastern shore of Halifax Harbour.

Sea Hurricane Ia, V7421, W-2D of 760 Naval Air Squadron, RNAS Yeovilton, Somerset, summer 1942 | No.760 Naval Air Squadron (NAS) originally formed at RNAS Eastleigh, Hampshire in April 1940, but moved to RNAS Yeovilton in the September as part of the Fleet Fighter School pool, an air combat training unit for Fleet Air Arm pilots. V7421 had already served with two frontline RAF Squadrons plus 52 OTU, before transferring to the FAA, and was a very well used machine by the time it had been 'navalised' and allocated to 760 NAS. Basically finished in the Temperate Sea Scheme, it shows signs of its previous schemes including most of the engine cowling in Dark Earth and Dark Green, and a very weathered finish, albeit enlivened by the large Popeye cartoon under cockpit.

Sea Hurricane Ib, Z4550, 6G, 880 Naval Air Squadron, HMS *Indomitable*, August 1942, Operation *Pedestal* | Based aboard HMS *Indomitable*, which was part of the escort to the Malta convoy code-named Operation *Pedestal*, 880 NAS operated Sea Hurricane Ibs, essentially navalised Hurricane Mk Is fitted with arrester hooks. Finished in the Temperate Sea Scheme with Sky under surfaces, Z4550 was one of the Sea Hurricanes involved in 'Pedestal' for which it had yellow identification markings applied to the wing leading edges. 880 NAS applied its aircrafts' individual identification letters on the fuselage aft of the roundel, with the full code on the inner wing leading edges. A black matchstick 'Saint' cartoon on a white background was carried under Z4550's cockpit on the starboard side, and a section of the nose cowling immediately behind the spinner was also painted white.

Hurricane IIb, BG766, Royal Naval Air Squadron Malta, Hal Far, Malta, late 1942/early 1943 | Although BG766 wasn't strictly a Sea Hurricane as such, it was operated by the Royal Naval Air Squadron Malta, based at Hal Far, which was an amalgamation of the Albacore and Swordfish units operating from Malta along with the survivors from the Fulmar Intruder Flight. There were originally four pilots from this unit flying ex-RAF Hurricanes that had been transferred over to the FAA, repainted in dark blue-grey upper surfaces with Azure Blue (or perhaps Sky Blue) under surfaces, on intruder operations and ASR patrols from Malta. The bomb racks were 'borrowed' from RAF Beaufighters.

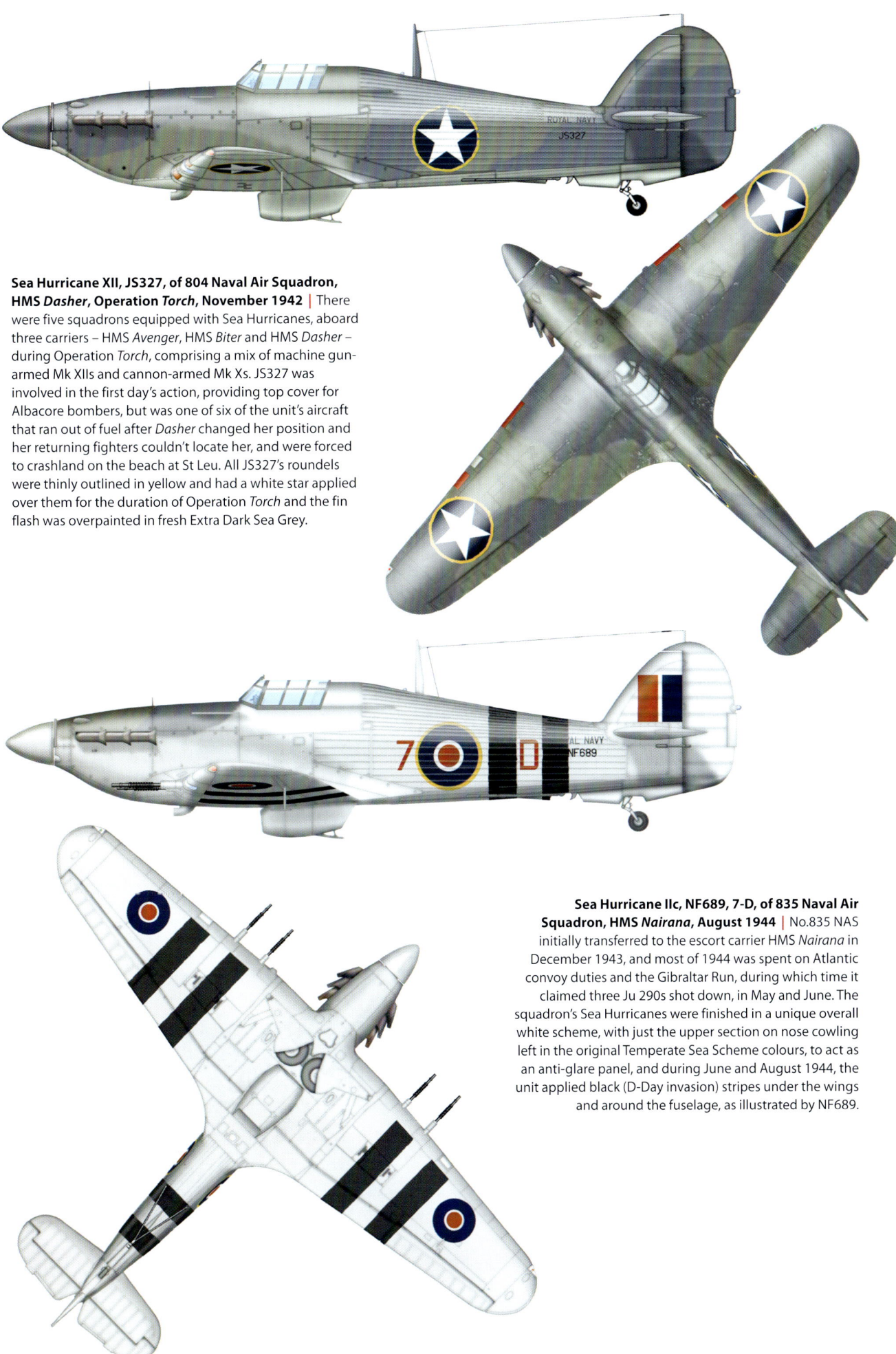

Sea Hurricane XII, JS327, of 804 Naval Air Squadron, HMS *Dasher*, Operation *Torch*, November 1942 | There were five squadrons equipped with Sea Hurricanes, aboard three carriers – HMS *Avenger*, HMS *Biter* and HMS *Dasher* – during Operation *Torch*, comprising a mix of machine gun-armed Mk XIIs and cannon-armed Mk Xs. JS327 was involved in the first day's action, providing top cover for Albacore bombers, but was one of six of the unit's aircraft that ran out of fuel after *Dasher* changed her position and her returning fighters couldn't locate her, and were forced to crashland on the beach at St Leu. All JS327's roundels were thinly outlined in yellow and had a white star applied over them for the duration of Operation *Torch* and the fin flash was overpainted in fresh Extra Dark Sea Grey.

Sea Hurricane IIc, NF689, 7-D, of 835 Naval Air Squadron, HMS *Nairana*, August 1944 | No.835 NAS initially transferred to the escort carrier HMS *Nairana* in December 1943, and most of 1944 was spent on Atlantic convoy duties and the Gibraltar Run, during which time it claimed three Ju 290s shot down, in May and June. The squadron's Sea Hurricanes were finished in a unique overall white scheme, with just the upper section on nose cowling left in the original Temperate Sea Scheme colours, to act as an anti-glare panel, and during June and August 1944, the unit applied black (D-Day invasion) stripes under the wings and around the fuselage, as illustrated by NF689.